National Plumbing
Codes Handbook

National Plumbing Codes Handbook

R. Dodge Woodson

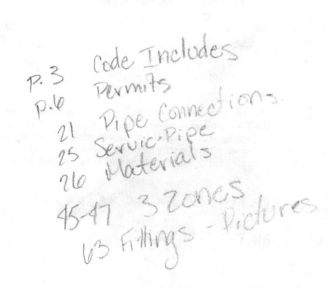

McGraw-Hill, Inc.

New York   San Francisco   Washington, D.C.   Auckland   Bogotá
Caracas   Lisbon   London   Madrid   Mexico City   Milan
Montreal   New Delhi   San Juan   Singapore
Sydney   Tokyo   Toronto

Library of Congress Cataloging-in-Publication Data

Woodson, R. Dodge (Roger Dodge), date.
   National plumbing codes handbook / by R. Dodge Woodson.
     p.   cm.
   Includes index.
   ISBN 0-07-071769-9
   1. Plumbing—Handbooks, manuals, etc. I. Title.
TH6125.W56   1993
696′.1—dc20   92-43491
                                                   CIP

1 2 3 4 5 6 7 8 9 0 DOC/DOC 9 9 8 7 6 5 4 3

ISBN 0-07-071769-9

The sponsoring editor for this book was Larry Hager, the editing supervisor was Nancy Young, and the production supervisor was Pamela A. Pelton. This book was set in Century Schoolbook. It was composed by North Market Street Graphics.

Printed and bound by R. R. Donnelley & Sons Company.

This book is dedicated to my loving daughter,
Afton Amber Woodson, and my supportive wife,
Kimberley Woodson.

Contents

Chapter 7. Fixtures

Chapter 8. Potable Water System

Chapter 9. After the Installation

Introduction

Congratulations; you are holding one of the most comprehensive, reader-friendly books available on the technical aspects of plumbing. If it is your desire to learn proper plumbing procedures, you have chosen the right book. This book is an indispensable tool for all licensed plumbers. It answers code questions on all major plumbing codes in a way no other book does. Whether you are sizing pipe or seeking a code interpretation, this book allows you to work faster, easier, and more profitably. If you are preparing for the plumbers licensing exam, this book will act as your private tutor. If you have no desire to earn a plumber's license but want to know how to perform plumbing like a professional, this book can show you how. You don't have to be a plumber to understand this book. Unlike plumbing codes that are written in cryptic language, this book is written so that the average reader will have no trouble understanding it.

If pictures are worth a thousand words, the word count in this book soars. There are many professional photographs between these pages that illustrate the text clearly. Line drawings, tables, and related illustrations are abundant. The combination of a conversational-style text and numerous illustrations come together to give you one of the best books available on technical plumbing procedures.

This book is an excellent companion to your local code book, and it offers special tips on studying for your licensing exam. There are even sample tests for you to take to assess your plumbing knowledge.

What makes this book so special? The author and his way with words. R. Dodge Woodson is a master plumber and gas-fitter, licensed in several states. Dodge has been involved with the plumbing trade for nearly 20 years. He has also instructed plumbing apprentice classes and code classes at the Central Maine Technical College. In addition to his plumbing prowess, Dodge is a prolific author of how-to books. When you combine the knowledge, experience, and skills Dodge brings to this book, you will be hard-pressed to find a better value.

Feel free to thumb through the pages. Look at the subheadings and illustrations. Read a paragraph here and there; you will see how easy Dodge makes it to learn a difficult trade. We are sure you will agree, this book is must-reading for anyone with a serious interest in plumbing.

How to Use This Book

This book is not a plumbing code book, and it will not replace your local code book, but it will help you to gain a better understanding of your local plumbing code.

There are three major plumbing codes currently being used in the United States. They are the Uniform Plumbing Code, the BOCA National Plumbing Code, and the Standard Plumbing Code. The Standard Plumbing Code is frequently called the Southern Code.

Different parts of the country adopt plumbing codes and frequently alter them to fit local needs. One of the three major codes is normally used as a boilerplate for local codes. There are instances when a local jurisdiction will adopt portions of multiple codes and combine them into a single local code.

In addition to the three major codes, there are other code examples available for local authorities to adopt. Considering the potential for so many possibilities in a local plumbing code, it would be nearly impossible to write a book that would address the requirements of each and every code.

Local Code Variations

Depending upon where you will be working, the plumbing code may have additional or different requirements from the procedures I describe. To be safe, you must always check and confirm local code requirements before doing any plumbing.

Local codes can change from town to town. I have worked in areas where a 15-minute drive would put you in a jurisdiction using a different code. With the possibility for so many variables, you must check local requirements in every jurisdiction.

This book provides good plumbing principles and procedures, as they are often outlined by various codes. In some cases, it is surprising how different two codes can be. This book is written with the use of a zone system. While it is not clear what every individual jurisdiction uses for an active plumbing code, there are statistics to show general code usages. Based on this information, I have broken the United States into zones for the various major plumbing codes.

Since there are three major plumbing codes, there are also three zones. The states in zone one are shown in Table I.1, states in zone two are shown in Table I.2, and zone three comprises the states listed in Table I.3.

As you read this book, you will notice references to the three zones. In many chapters you will find reminder notes. These reminder notes are provided to give you quick reference for significant differences between the codes used in different zones.

I wish I could give you concrete, clear-cut instructions for all plumbing codes, but I cannot. Plumbing codes are frequently amended, and successful books can have long shelf lives. These two reasons make it difficult to give you hard-and-fast rules to follow. However, I can and will give you the basics for proper plumbing procedures and practices. Always check with your local code enforcement office before doing any plumbing. I have made this book as accurate as I could, but people do make mistakes, and the production of a book can provide opportunity for mistakes. If you use this book as your guide, and check with local authorities before attempting any work, you will be fine.

TABLE I.1   States in Zone One

Washington
Oregon
California
Nevada
Idaho
Montana
Wyoming
North Dakota
South Dakota
Minnesota
Iowa
Nebraska
Kansas
Utah
Arizona
Colorado
New Mexico
Indiana
Parts of Texas

TABLE I.2   States in Zone Two

Alabama
Arkansas
Louisiana
Tennessee
North Carolina
Mississippi
Georgia
Florida
South Carolina
Parts of Texas
Parts of Maryland
Parts of Delaware
Parts of Oklahoma
Parts of West Virginia

TABLE I.3   States in Zone Three

Virginia
Kentucky
Missouri
Illinois
Michigan
Ohio
Pennsylvania
New York
Connecticut
Massachusetts
Vermont
New Hampshire
Rhode Island
New Jersey
Parts of Delaware
Parts of West Virginia
Parts of Maine
Parts of Maryland
Parts of Oklahoma

Acknowledgments

I would like to acknowledge and thank the following sources of illustrative materials used in this book:

CR/PL, Inc.
Crane Plumbing Fixtures
2020 Dempster Plaza Suite 1235
Evanston, Illinois 60202

Ridge Tool Company
400 Clark St.
P. O. Box 4023
Elyria, Ohio 44036-2023

Southern Building Code Congress International
900 Montclair Road
Birmingham, Alabama 35213-1206

Universal-Rundel Corporation
303 North Street
New Castle, Pennsylvania 16103

Vanguard Plastics, Inc.
831 N. Vanguard St.
McPherson, Kansas 67460

Administrative Policies and Procedures

Administrative policies and procedures are what make the plumbing code effective. Without the proper procedures and administration, the plumbing code would be little more than an organized outline for good plumbing procedures. To be effective, the code must be enforced. To be fair, the rules for the administration of the code must be clear to all who work with it. Administrative policies dictate the procedure for code enforcement, interpretation, and implementation.

There is a fine line between administration and enforcement. This chapter deals with administration. Chapter 2 addresses the issue of code enforcement. While the two matters are broken into two separate chapters, there will be some overlap of material. Administration and enforcement are so closely related it is necessary to commingle the two from time to time.

The rules, regulations, and laws used to make the plumbing code are structured around facts. These facts are the result of research into ways to protect the health of our nation. As a plumber, you are responsible for the health and sanitation of the public. A mistake or a code violation could result in widespread illness or even death.

In some jurisdictions the plumbing code comprises rules. In other areas it is a compilation of laws. There is a big difference between a rule and a law. When you are working with a rule-based code, you are subject to various means of punishment for violating the rules. The punishment may mean the suspension or revocation of your license. There may be cash fines required for violations, but there is no jail time. For jurisdictions using a law-based code you could find yourself behind bars for violating the plumbing code.

The procedure required to obtain a plumber's license is not easy. Some people feel there are too many restraints in the licensing requirements for plumbers, but these people are not aware of the heavy responsibility plumbers must bear. The public often perceives a plumber as someone who works in sewers, has a poor education, and is slightly more than a common laborer. This is a false perception.

Professional plumbers are much more than sewer rats. Today's plumbers are generally well educated and have the ability to perform highly technical work. The mathematical demands on a plumber could perplex many well-educated people. The mechanical and physical abilities of plumbers are often outstanding.

While drain cleaning is a part of the trade, so is the installation of $2500 gold faucets. The plumbing trade is not all sewage and water. Plumbers are known for their fabled high incomes, and it is true that good plumbers make more money than many white collar professionals. The trade can offer a prosperous living, but it must be worked for.

Whether you are a plumber or an apprentice, you have much to be proud of. The plumbing trade is more than a job. As a professional plumber you will have the satisfaction of knowing you are helping to maintain the health of the nation and the integrity of our natural resources.

It was not long ago when there was no plumbing code or code enforcement. People could pollute our lakes and streams with their ineffective cesspools and outhouses. The plumbing code is designed to stop pollution and health hazards. The code is part of any plumber's life and career. Learning the code can be a laborious task, but the self-satisfaction obtained when you master the code is well worth the effort.

When you earn your master's license, you will have the opportunity to run your own shop. Having your own plumbing business can be quite profitable. To realize these profits you must first learn the code and the trade. Then, you must pass the tests for your journeyman and master's licenses. During your learning stages you are earning a good wage and providing a vital service to the community.

There are few professions that allow you to earn a good living while you are gaining the skills necessary to master your craft. When you attend college, you must pay for your education. With plumbing, you get paid to learn. After your training you are unlimited in the wealth you can build from your own business.

The first step toward financial independence as a plumber is a clear understanding of the plumbing code. Unlicensed plumbers are not allowed to work in many jurisdictions. Your license is your ticket to respectable paychecks and a solid future. Now, let's see how the administrative policies and procedures for the plumbing code will affect you.

What Does the Plumbing Code Include?

The plumbing code includes all major aspects of plumbing installations and alterations. Design methods and installation procedures are a cornerstone of the plumbing code. Sanitary piping for the disposal of waste, water, and sewage is controlled by the code. Potable water supplies fall under the jurisdiction of the plumbing code. Storm water, gas piping, chilled-water piping, hot-water piping, and fire sprinklers are all dealt with in the plumbing, mechanical, and building codes.

The plumbing code is meant to ensure the proper design and installation of plumbing systems and to ensure public health and safety. It is intended to be interpreted by the local code enforcement officer. The interpretation of the code officer may not be the same as yours, but it is the code officer's option to determine the meaning of the code under questionable circumstances.

How the Code Pertains to Existing Plumbing

The plumbing code requires any alterations or repairs to an existing plumbing system to conform to the general regulations of the code, as they would apply to new installations. No alteration or repair shall cause an existing plumbing system to become unsafe. Further, the alterations or repairs shall not be allowed to have a detrimental effect on the operation of the existing system.

For example, if a plumber is altering an existing system to add new plumbing, the plumber must make all alterations in compliance with code requirements. It would be a violation of the code to add new plumbing to a system that was not sized to handle the additional load of the increased plumbing.

There are provisions in the codes to allow existing conditions that are in violation of the current code to be used legally. If an existing condition was of an approved type prior to the current code requirements, that existing condition may be allowed to continue in operation so long as it is not creating a safety or health hazard.

If the use or occupancy of a structure is being changed, the change must be approved by the proper authorities. It is a violation of the plumbing code to change the use or occupancy of a structure without the proper approvals. For example, it would be a breach of the code to convert a residential dwelling to a professional building without the approval of the code enforcement office.

Small Repairs

Small repairs and minor replacements of existing plumbing may be made without bringing the entire system into compliance with the current plumbing code standards. These changes must be made in a safe and sanitary method and must be approved. For example, it would be permissible to repair a leak in a ½-inch (in) pipe without changing the ½-in pipe to a ¾-in pipe, even if the current code required a ¾-in pipe under the present use. It would also be allowed to replace a defective S-trap with a new S-trap, even though S-traps are not in compliance with the current code requirements. In general, if you are only doing minor repair or maintenance work, you are not required to update the present plumbing conditions to current code requirements.

It is incumbent upon the owner of a property to keep the plumbing system in good and safe repair. The owner may designate an agent to assume responsibility for the condition of the plumbing, but it is mandatory that the plumbing be kept safe and sanitary at all times.

Relocation and Demolition of Existing Structures

If a building is moved to a new location, the building's plumbing must conform to the current code requirements of the jurisdiction where the structure is located. In the event a structure is to be demolished, it is the owner's, or the owner's designated agent's, responsibility to notify all companies, persons, and entities having utilities connected to the structure. These utilities may include, but are not limited to, water, sewer, gas, and electrical connections.

Before the building can be demolished or moved, the utilities having connections to the property must disconnect and seal their connections in an approved manner. This applies to water meters and sewer connections, as well as other utilities.

Materials

All materials used in a plumbing system must be approved for use by the code enforcement office. These materials shall be installed in accordance with the requirements of the local code authority. The local code officer has the authority to alter the provisions of the plumbing code so long as the health, safety, and welfare of the public is not endangered.

A property owner, or that owner's agent, may request a variance from the standard code requirements when conditions warrant a hardship. It is up to the code officer to decide whether or not the variance should be granted. The application for a variance and the final decision of the code officer shall be in writing and filed with the code enforcement office.

The use of previously used materials shall be open to the discretion of the local code officer. If the used materials have been reconditioned and tested and are in working condition, the code officer may allow their use in a new plumbing system.

Alternative materials and methods not specifically identified in the plumbing code may be allowed under certain circumstances. If the alternatives are equal to the standards set forth in the code for quality, effectiveness, strength, durability, safety, and fire resistance, the code officer may approve the use of the alternative materials or methods.

Before alternative materials or methods are allowed for use, the code officer can require adequate proof of the properties of the materials or methods. Any costs involved in testing or providing technical data to substantiate the use of alternative materials or methods shall be the responsibility of the permit applicant.

Code Officers

Code officers are responsible for the administration and enforcement of the plumbing code. They are appointed by the executive authority for the community. Code officers may not be held liable on a personal basis when working for a jurisdiction. Legal suits brought forth against code officers, arising from on-the-job disputes, will be defended by the legal representative for the jurisdiction.

The primary function of code officers is to enforce the code. Code officers are also responsible for answering questions pertinent to the materials and installation procedures used in plumbing. When application is made for a plumbing permit, the code officer is the individual who receives the application. After reviewing a permit application, the code officer will issue or deny a permit.

Once a permit is issued by the code officer, it is the code officer's duty to inspect all work to ensure it is in compliance with the plumbing code. When code officers inspect a job, they are looking for more than just plumbing. These inspectors will be checking for illegal or unsafe conditions on the job site. If the safety conditions on the site or the plumbing are found to be in violation of the code, the code officer will issue a notice to the responsible party.

Code officers normally perform routine inspections personally. However, inspections may be performed by authoritative and recognized services or individuals instead of the code officers. The results of all inspections shall be documented in writing and certified by an approved individual.

If there is ever any doubt as to the identity of a code officer, you may request to see the inspector's identification. Code officers are required to carry official credentials while discharging their duties.

Another aspect of the code officer's job is the maintenance of proper records. Code officers must maintain a file of all applications, permits, certificates, inspection reports, notices, orders, and fees. These records are required to be maintained for as long as the structure they apply to is still standing, unless otherwise stated in other rules and regulations.

Plumbing Permits

Most plumbing work, other than minor repairs and maintenance, requires a permit. These permits must be obtained prior to the commencement of any plumbing work. The code enforcement office provides forms to individuals wishing to apply for plumbing permits. The application forms must be properly completed and submitted to the code enforcement officer.

The permit application shall give a full description of the plumbing to be done. This description must include the number and type of plumbing fixtures to be installed. The location where the work will be done and the use of the structure housing the plumbing must also be disclosed.

The code officer may require a detailed set of plans and specifications for the work to be completed. Duplicate sets of the plans and specifications may be required so that copies can be placed on file in the code enforcement office. If the description of the work deviates from the plans and specifications submitted with the permit application, it may be necessary to apply for a supplementary permit.

The supplementary permit will be issued after a revised set of plans and specifications have been given to the code officer and approved. The revised plans and specifications must show all changes in the plumbing that are not in keeping with the original plans and specifications.

Plans and specifications may not be required for the issuance of a plumbing permit. However, if plans and specifications are required, they may require a riser diagram and a general blueprint of the structure. The riser diagram must be very detailed. The diagram must indicate pipe size, direction of flow, elevations, fixture-unit ratings for drainage piping, horizontal pipe grading, and fixture-unit ratings for the water distribution system.

If the plumbing to be installed is an engineered system, the code officer may require details on computations, plumbing procedures, and other technical data. Any application for a plumbing permit to install new plumbing might require a site plan. The site plan must identify the locations of the water service and sewer connections. The location of all vent stacks and their proximity to windows or other ventilation openings must be shown.

In the event new plumbing is being installed in a structure served by a private sewage disposal system, there are yet more details to be included in the site plan. When a private sewage system is used, the plan must show the location of the system and all technical information pertaining to the proper operation of the system.

When a plumbing permit is applied for, the code officer will process the application in a timely manner. If the application is not approved, the code officer will notify the applicant, in writing, of the reasons for denial. If the applicant fails to follow through with the issuance of a permit within 6 months from the date of application, the permit request can be considered void. Once a permit is issued, it may not be assigned to another person or entity. Permits will not be issued until the appropriate fees are paid. Fees for plumbing permits are provided by the local jurisdiction.

Plumbing permits bear the signature of the code officer or an authorized representative. The plans submitted with a permit application will be labeled as approved plans by the code officer. One set of the plans will be retained by the code enforcement office. A set of approved plans must be kept on the job site. The approved plans kept on the job must be available to the code officer or an authorized representative for inspection at all reasonable times.

It is possible to obtain permission to begin work on part of a plumbing system before the entire system has been approved. For example, you might be given permission to install the underground plumbing for a building before the entire plumbing system is approved. These partial permits are issued by the code officer with no guarantee the remainder of the work will be approved. If you proceed to install the partial plumbing, you do so at your own risk in regard to the remainder of the plumbing not yet approved.

There are time limits involved after a plumbing permit is issued. If work is not started within 6 months of the date a permit is issued, the permit may become void. If work is started, but then stalled or abandoned for a period of 6 months, the permit may be rendered useless. A permit may be revoked if the code officer finds that the permit was issued under false information. Misrepresentation in the application for a permit or on the plans submitted for review is reason for the revocation of a plumbing permit.

All work performed must be done according to the plans and specifications submitted to the code officer in the permit application process. All work must be in compliance with the plumbing code. Code officers are required to conduct inspections of the plumbing being installed during the installation and upon completion of the installation. The details of these inspections are described in Chapter 2.

Multiple Plumbing Codes

There are many plumbing codes in use. Each local jurisdiction generally takes an existing code and amends it to local needs. There are three primary plumbing codes, the codes for zones one, two, and three.

These three codes are similar in many ways and very different in others. In addition to the three primary codes, there are other codes in existence. To be sure of your local code requirements, you must check with the local code enforcement office.

This book is written to explain good plumbing procedures. However, various jurisdictions have different opinions of what good plumbing procedures are. Some states, counties, or towns adapt an existing code without much revision. Other areas make significant changes in the established code that is used as a model. It would not be unheard of to find a jurisdiction working with regulations from multiple plumbing codes. In light of these facts, always check with your local authorities before performing plumbing work.

2

Regulations, Permits, and Code Enforcement

When working with the plumbing code, you must be aware of the requirements and procedures involved with regulations, permits, and enforcement. These three elements work together to ensure the proper use of the plumbing code. This chapter is going to teach you about all three key elements, but it will also do much more. You are going to learn the facts about pipe protection, health, safety, pipe connections, temporary toilet facilities, and more. There will even be tips on how to work with plumbing inspectors, instead of against them.

Regulations

What are regulations? They are the rules or laws used to regulate activity, in this case, the performance of plumbing. The plumbing code is governed by rules in some states and laws in others. If you violate the regulations in a rule-base state, you may face disciplinary action, but not jail. In states where the plumbing regulations are laws, you risk going to jail, if you violate the regulations. It may be difficult to imagine going to jail for violating a plumbing regulation, but it could happen. Certain violations could result in personal injury or death. To protect yourself, and others, it is important to understand and abide by the regulations governing the plumbing trade.

Existing conditions

The first regulations we are going to examine have to do with existing conditions. This is an area where many people have difficulty in determining their responsibility to the plumbing code. Generally, any existing condition that is not a hazard to health and safety is allowed to remain in existence. However, when existing plumbing is altered, it may have to be brought up to current code requirements.

While the code is normally based on new installations, it does apply to existing plumbing that is being altered. These alterations may include repairs, renovations, maintenance, replacement, and additions. The question of when work on existing plumbing must meet code requirements is one that plagues many plumbers. Let's address this question and clear it up.

The code is only concerned with changes being made to existing plumbing. As long as the existing plumbing is not creating a safety or health hazard, and is not being altered, it does not fall under the scrutiny of most plumbing codes. If you are altering an existing system, the alterations must comply with the code requirements, but there may be exceptions to this rule. For example, if you were replacing a kitchen sink and there was no vent on the sink's drainage, code would require you to vent the fixture. Where undue hardship exists in bringing an existing system into compliance, the code officer may grant a variance.

In the case of the kitchen sink replacement, such a variance may be in the form of permission to use a mechanical vent (Fig. 2.1). Whenever you encounter a severe hardship in making old plumbing come up to

Figure 2.1 Mechanical vent.

code, talk with your local code officer. The code officer should be able to offer some form of assistance, either in the form of a variance or as advice on how to accomplish your goal.

Since the code does come into play with repairs, maintenance, replacement, alterations, and additions, let's see how it affects each of these areas. If you are repairing a plumbing system, you must be aware of code requirements. If no health or safety hazards exist, non-conforming plumbing may be repaired to keep it in service.

Do you need to apply for a plumbing permit to replace a faucet? No, faucet replacement does not require a permit, but it does require the replacement to be made with approved materials and in an approved manner. Remember this, you do need a permit to replace a water heater. Even if the replacement heater is going in the same location and connecting to the same existing connections, you must apply for a permit and have your work inspected. An improperly installed water heater can become a serious hazard, capable of causing death and destruction. Your failure to comply with the code in these circum-stances could ruin your life and the lives of others.

Routine maintenance of a plumbing system must be done according to the code, but it does not require a permit. Alterations to an existing system may require the issuance of a permit, depending upon the nature of the alteration. In any case, alterations must be done with approved materials and in an approved manner. When adding to a plumbing system, you will normally need to apply for a permit. Adding new plumbing will come under the authority of the plumbing code and will generally require an inspection.

You can get yourself into deep water when adding to an existing sys-tem. When you add new plumbing to an old system, you must be concerned with the size and ability of the old plumbing. Increasing the number of fixture units entering an old pipe may force you to increase the size of the old pipe. This can be very expensive, especially when the old pipe happens to be the building drain or sewer. Before you install any new plumbing on an old system, verify the size and ratings of the old system. If your new work will rely on the old plumbing to work, the old plumbing must meet current code requirements.

Beware of the change-in-use regulations. If you will be performing commercial plumbing, this regulation can have a particularly serious effect on your plumbing costs and methods. If the use of a building is changed, the plumbing may also have to be changed. The change-in-use regulations come into play most often on commercial properties, but they could affect a residential building.

Assume for a moment that you receive a request to install a three-bay sink in a convenience store. You discover that the store's owner is having the sink installed in order to perform food preparation for a new deli in the store. This store has never been equipped for food preparation and service. What complications could arise from this situation? First, zoning may not allow the store to have a deli. Second, if the use is allowed, the plumbing requirements for the store may soar. There could be a need for grease traps, indirect wastes, and a number of other possibilities. When you are asked to perform plumbing that involves a change of use, investigate your requirements before committing to the job.

The remaining general regulations of the plumbing code are easily understood. Many of them were covered in the last chapter. By reading your local code book you should have no trouble in understanding the regulations. Now, let's move on to permits.

Permits

Permits are generally required for various plumbing jobs. When a permit is required, it must be obtained before any work is started. Minor repairs and drain cleaning do not require the issuance of permits. Permits were discussed in the last chapter, but we will go over them again and add some to what you have already learned.

In most cases plumbing permits can only be obtained by master plumbers or their agents. In some cases, however, homeowners may be allowed to receive plumbing permits for work to be done by themselves on their own homes. Permits are obtained from the local code enforcement office and that office provides the necessary forms for permit applications.

The information required to obtain a permit will vary from jurisdiction to jurisdiction. You may be required to submit plans, riser diagrams (Figs. 2.2 and 2.3), and specifications for the work to be performed. At a minimum, you will likely be required to adequately describe the scope of work to be performed, the location of the work, the use of the property, and the number and type of fixtures being installed.

The degree of information required to obtain a permit is determined by the local code officer. It is not unusual for the code officer to require two sets of plans and specifications for the work to be performed. The detail of the plans and specifications is also left up to the judgment of the code officer. Requirements may include details of pipe sizing, grade, fixture units, and any other information the code officer may deem pertinent.

If your work will involve working with a sewer or water service, expect to be asked for a site plan. The site plan should show the locations of the water service and sewer (Fig. 2.4). If you will be working with a private sewage disposal system, its location should be indicated

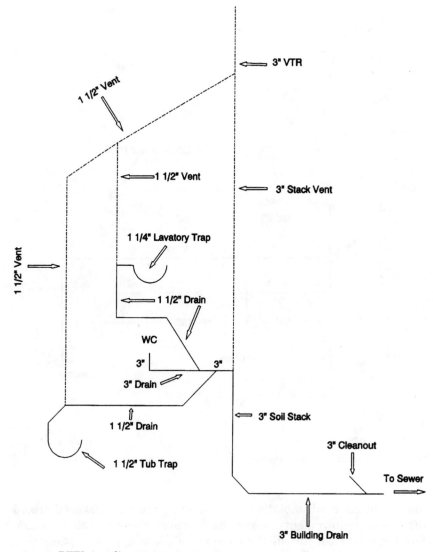

Figure 2.2 DWV riser diagram.

on the site plan (Fig. 2.5). Once your plans are approved, any future changes to the plans must be submitted to and approved by the code officer.

Plumbing permits must be signed by the code officer or an authorized representative. If you submitted plans with your permit application, the plans will be labeled with appropriate wording to prove they have been reviewed and approved. If it is later found that the approved

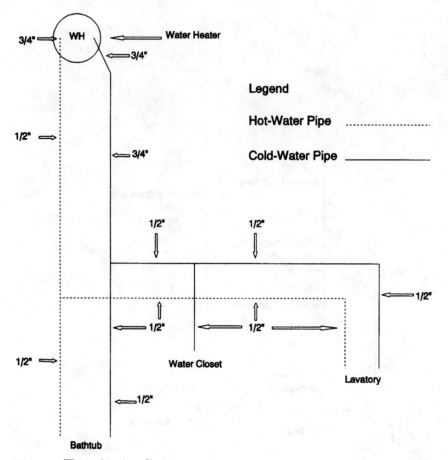

Figure 2.3 Water pipe riser diagram.

plans contain a code violation, the plumbing must be installed according to code requirements even if the approved plans contain a nonconforming use. Most jurisdictions require a set of approved plans to be kept on the job site and available to the code officer at all reasonable times.

If plans are required, they must be approved before a permit is issued. All fees associated with the permit must be paid prior to the issuance of the permit. These fees are established by local jurisdictions. After a permit is issued, all work must be done in the manner presented during the permit application process.

It is possible to obtain a partial permit. This is a permit that approves a portion of work proposed for completion. When time is important, it may be possible to obtain these partial permits, but there

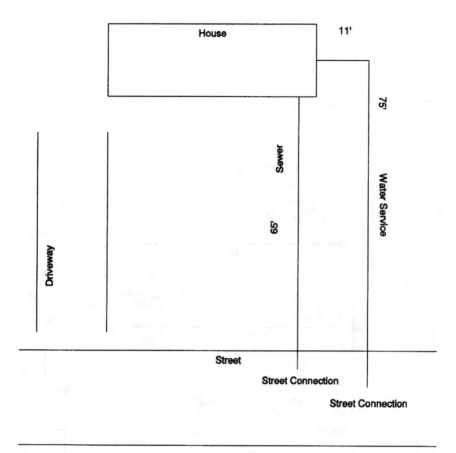

Figure 2.4 Site plan with city water and sewer.

is risk involved. Assume you obtained permission and a permit to install your underground plumbing but had not yet been issued a permit for the remainder of the plumbing. With freezing temperatures coming on, you decide to install your groundworks so that the concrete floor can be poured over the plumbing before freezing conditions arrive. This is a good example of how and why partial approvals are good, but look at what could happen.

You have installed your underground plumbing and the concrete is poured. After awhile you are notified by the code officer that the proposed above-grade plumbing is not in acceptable form and will require major changes. These changes will affect the location and size of your underground plumbing. What do you do now? Well, you are probably going to spend some time with a jackhammer or concrete saw. The underground plumbing must be changed, or the above-ground plumb-

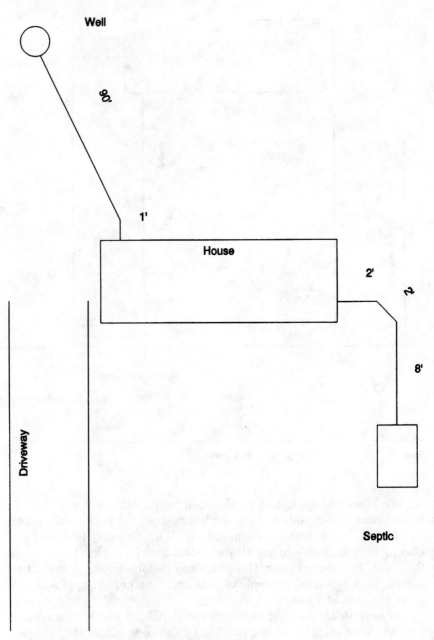

Figure 2.5 Site plan with private water and sewer.

ing must be redesigned to work with the groundworks. In either case, you have trouble and expense that would have been avoided if you had not acted on a partial approval. Partial approvals have their place, but use them cautiously.

How can a plumbing permit become void? If you do not begin work within a specified time, normally 6 months, your permit will be considered abandoned. When this happens, you must start the entire process over again to obtain a new permit. Plumbing permits can be revoked by the code official. If it is found that facts given during the permit application were false, the permit may be revoked. If work stops for an extended period of time, normally 6 months, a permit may be suspended.

Code Enforcement

Code enforcement is generally performed on the local level, by code enforcement officers. These individuals are frequently referred to as inspectors. It is their job to interpret and enforce the regulations of the plumbing code. Since code enforcement officers have the duty of interpreting the code, there may be times when a decision is reached that appears to contradict the code book. The code book is a guide, not the last word. The last word comes from the code enforcement officer. This is an important fact to remember. Regardless of how you interpret the code, it is the code officer's decision that is final.

Inspections

Every job that requires a permit also requires inspection. Many jobs require more than one inspection. In the plumbing of a new home there may be as many as four inspections. One inspection would be for the sewer and water-service installation. Another inspection might be for underground plumbing. Then you would have a rough-in inspection for the pipes that are to be concealed in walls and ceilings. Then, when the job is done, there will be a final inspection.

These inspections must be done while the plumbing work is visible. A test of the system is generally required, with pressure from either air or water. Normally the inspection is done by the local code officer, but not always. The code officer may accept the findings of an independent inspection service. Before independent inspection results will be accepted, the inspection service must be approved by the code officer. Independent inspection services are commonly used to inspect prefabricated construction.

Plumbing inspectors are generally allowed the freedom to inspect plumbing at any time during normal business hours. These inspectors cannot enter a property without permission unless they obtain a search warrant or other proper legal authority. Permission for entry is frequently granted by the permit applicant when the permit is signed.

What inspectors look for

When plumbing inspectors look at a job, they are looking at many aspects of the plumbing. They will inspect to see that the work is installed in compliance with the code and in a way that the plumbing will be likely to last for its normal lifetime. Inspectors will check to see that all piping is tested properly and that all plumbing is in good working order.

What powers do plumbing inspectors have?

Plumbing inspectors can be considered the plumbing police. These inspectors have extreme authority over any plumbing-related issue. If plumbing is found to be in violation of the code, plumbing inspectors may take several forms of action to rectify the situation.

Normally, inspectors will advise the permit holder of the code violations and allow a reasonable time for correction of the violations. This advice will come in the form of written documents and will be recorded in an official file. If the violations are not corrected, the code officer will take further steps. Legal counsel may be consulted. After a legal determination is made, action may be taken against the permit holder in violation of the plumbing code. This could involve cash fines, license suspension, license revocation, and in extreme cases, jail.

Code officers have the power to issue a stop work order. This order requires all work to stop until code violations are corrected. These orders are not used casually; they are used when an immediate danger is present or possible. Code officers do have a protocol to follow in the issuance of stop work orders. If you ever encounter a stop work order, stop working. These orders are serious and violation of a stop work order can deliver more trouble than you ever imagined.

When code officers inspect a plumbing system and find it to be satisfactory, they will issue an approval on the system. This allows the pipes to be concealed and the system to be placed into operation. In certain circumstances code officers may issue temporary approvals. These temporary approvals are issued for portions of a plumbing system when conditions warrant them.

When a severe hazard exists, plumbing inspectors have the power to condemn property and force occupants to vacate it. This power would only be used under extreme conditions, where a health or safety hazard was present.

What can you do to change
a code officer's decision?

If you feel you have received an unfair ruling from a code officer, you may make a formal request to have the decision changed. In doing this, you must make your request to an appeal board. Your reasons for an appeal must be valid and must pertain to specific code requirements. Your appeal could be based on what you feel is an incorrect interpretation of the code. If you feel the code does not apply to your case, you have reason for an appeal. There are other reasons for appeal, but you must specify why the appeal is necessary and how the decision you are appealing is incorrect.

Tips on Health and Safety

Health and safety are two key issues in the plumbing code. These two issues are, by and large, the reasons for the plumbing code's existence. The plumbing code is designed to assure health and safety to the public. Public health can be endangered by faulty or improperly installed plumbing. Code officers have the power and duty to condemn a property where severe health or safety risks exist. It is up to the owner of each property to maintain the plumbing in a safe and sanitary manner.

When it comes to safety, there are many more considerations than just plumbing pipes. Most safety concerns arise in conjunction with plumbing but not from the plumbing itself. It is far more likely that a safety hazard will result from the activities of a plumber on other aspects of a building. An example could be cutting so much of a bearing timber that the structure becomes unsafe. Perhaps a plumber removes a wire from an electric water heater and leaves it exposed and unattended; this could result in a fatal shock to someone. The list of potential safety risks could go on for pages, but you get the idea. It is your responsibility to maintain safe and sanitary conditions at all times.

A part of maintaining sanitary conditions includes the use of temporary toilet facilities on job sites. It is not unusual for the plumbing code to require toilet facilities to be available to workers during the construction of buildings. These facilities can be temporary, but they must be sanitary and available.

Pipe Protection

It is the plumber's responsibility to protect plumbing pipes. This protection can take many forms. Here we are going to look at the basics of pipe protection. You will gain insight into pipe protection needs that you may have never considered before.

Backfilling

When backfilling over a pipe, you must take measures to prevent damage to the pipe. The damage can come in two forms, immediate and long-term. If you are backfilling with material that contains large rocks or other foreign objects, the weight or shape of the rocks and objects may puncture or break the pipe. The long-term effect of having large rocks next to a pipe could result in stress breaks. It is important to use only clean backfill material when backfilling pipe trenches.

Even the weight of a large load of backfill material could damage the pipe or its joints. Backfill material should be added gradually. Layers of backfill between 6 and 12 inches deep are typically recommended as the average interval for filling a trench. Each layer of this backfill should be compacted before the next load is dumped.

Flood protection

If a plumbing installation is made in an area subject to flooding, special precautions must be taken. High water levels can float pipes and erode the earth around them. If your installation is in a flood area, consult your local code officer for the proper procedures in protecting your pipes.

Penetrating an exterior wall

When a pipe penetrates an exterior wall, it must pass through a sleeve. The sleeve should be at least two pipe sizes larger than the pipe passing through it. Once the pipe is installed, the open space between the pipe and the sleeve should be sealed with a flexible sealant. By caulking around the pipe you eliminate the invasion of water and rodents.

Freezing

Pipes must be protected against freezing conditions. Outside, this means placing the pipe deep enough in the ground to avoid freezing. The depth will vary with geographic locations, but your local code officer can provide you with minimum depths. Above-ground pipes, in unheated areas, must be protected with insulation or other means of protection from freezing.

Corrosion

Pipes that tend to be affected by corrosion must be protected. This protection can take the form of a sleeve or a special coating applied to the pipe. For example, copper pipe can have a bad reaction when placed in contact with concrete. If you have a copper pipe extending through concrete, protect it with a sleeve. The sleeve can be a foam insulation or some other type of noncorrosive material. Some soils are capable of corroding pipes. If corrosive soil is suspected, you may have to protect entire sections of underground piping.

Pipe Connections

Pipe connections can require a variety of adapters when combining pipes of different types. It is important to use the proper methods when making any connections, especially when you are mating different types of pipe together. There are many universal connectors available to plumbers today. These special couplings are allowed to connect a wide range of various materials.

Male and female adapters have long been an acceptable method of joining opposing materials, but today the options are much greater. You can use compression fittings and rubber couplings to match many types of materials to each other. Special insert adapters allow the use of plastic pipe with bell-and-spigot cast iron.

Working with the System Instead of against It

Code officers are expected to enforce the regulations set forth by the code. Plumbers are expected to work within the parameters of the code. Naturally, plumbers and code officers will come into contact with each other on a regular basis. This contact can lead to some disruptive actions.

The plumbing code is in place to help people, not to hurt them. It is not meant to ruin your business or to place you under undue hardship in earning a living. It really is no different than our traffic laws. The traffic laws are there to protect all of us, but some people resent them. Some plumbers resent the plumbing code. They view it as a vehicle for the local jurisdiction to make more money while they, the plumbers, are forced into positions to possibly make less money.

When you learn to understand the plumbing code and its purpose, you will learn to respect it. You should respect it; it shows the importance of your position as a plumber to the health of our entire nation. Whether you agree with the code or not, you must work within the constraints of it. This means working with the inspectors.

When inspectors choose to play hardball, they hold most of the cards. If you develop an attitude problem, you may be paying for it for years to come. Even if you know you are right on an issue, give the inspector a place to escape; nobody enjoys being ridiculed.

The plumbing code is largely a matter of interpretation. If you have questions, ask your code officer for help. Code officers are generally more than willing to give advice. It is only when you walk into their offices with a chip on your shoulder that you are likely to hit the bureaucratic wall. Like it or not, you must learn to comply with the plumbing code and to work with code officers. The sooner you learn to work with them on amicable terms, the better off you will be.

Little things can mean a lot. Apply for your permit early. This eliminates the need to hound the inspector to approve your plans and issue your permit. Many jurisdictions require at least a 24-hour (h) advance notice for an inspection request, but even if your jurisdiction doesn't have this rule, be considerate, and plan your inspections in advance. By making life easier for the inspector, you will be helping yourself.

Chapter

3

Approved Materials and Their Connection

In order to do plumbing in the best and most cost-effective manner, you must choose the proper materials. Today's plumbers have so many materials to choose from that the decision of which material is best suited to the job can become perplexing. Take the water-distribution system as an example. When you are trying to decide what type of pipe to run your potable water through, you will have many options available. You could choose polybutylene (PB), chlorinated polyvinyl chloride (CPVC), or copper, just to name a few. If you decide to use copper, you must determine which copper to use. Will you use type M, type L, or type K? Can you use polyethylene (PE) or polyvinyl chloride (PVC)? As you can see, the choices can be confusing.

This chapter is going to give you a tour of the materials approved for use and the uses they are approved for. A material that is approved for above-ground use may not be allowed below ground. A pipe suitable for cold water may not work with hot water. This chapter will explain the different materials and give suggestions about which materials are best suited for specific uses. In addition, you will learn about approved connections.

Choosing the proper materials will help you in several ways. By working with only approved materials, you will not have to do the same job twice. Failure to use approved materials can result in code officers requiring you to rip out work already installed so that it can be replaced with proper materials. Effective material selection can save you money or make your bid for a job more competitive. By assessing all circumstances surrounding a job, your material selection can help you avoid problems later with callbacks and warranty work. Now, let's get down to business.

What Is an Approved Material?

An approved material is a material approved for specific uses, as determined by the local code enforcement office. Many approved materials and their approved uses are described in plumbing code books, but not all approved materials are listed. The materials detailed in code books represent the most commonly encountered approved materials. However, in certain cases other materials may be approved for use. This is frequently the case when a material exceeds the requirements listed in a code book.

The standards set in code books are normally the minimum acceptable standards. A product that exceeds these minimums may be allowed for use but not mentioned in the code. For general use, the materials listed in code books will be sufficient for your needs. Approved materials should be marked in a manner to identify themselves. Normally this identification will take the form of an embossed, molded, or indelible marking.

The type of identification marking used is frequently determined by the type of material being identified. Brass and copper fittings are often stamped. Plastic fittings usually have a molded marking. Pipe will typically carry a colored stripe and indelible letters to identify itself. For example, type M copper will have an indelible red stripe. Type L copper will have a blue stripe, and type K copper will have a green stripe. A yellow stripe will indicate a drain waste and vent (DWV) copper. By glancing at these color-codes, a code enforcement officer can quickly identify the type of copper being worked with.

A material is subject to local approval. For example, the water quality in a certain location may have an adverse effect on a particular type of pipe. In such a case, the local code enforcement office may deny the use of a material identified as an approved material in the code book. A similar deviation from the code may be made for pipe used below grade. The soils in some areas are not compatible with certain materials. Adjustments for local conditions may be made by the local authorities and must be checked to ensure the legality of material usage.

Materials approved for carrying potable water must not have a lead content of more than 8 percent. This applies to pipe and fittings. Solder used to join pipe and fittings for potable use may not contain more than 0.2 percent lead. There are several brands of lead-free solder available. The 50/50 solder, once common to the industry, is no longer approved for potable water systems.

When materials are approved, they are often approved for specific uses. It is not enough to say that PE pipe is approved for potable water use. In a sense, this is a true statement, but it has exceptions. PE pipe may not be used to convey hot water; it is not rated to convey hot water.

Hot water is water with a temperature of at least 110°F. However, PE is approved to carry potable water as a water-service pipe. So, you see, PE is approved for potable water use, but its range of use is limited. This type of regulation confuses many people, but this chapter will clear the confusion for you. Now, let's look at approved materials and their allowed uses.

Water-Service Pipe

A water-service pipe is a pipe extending from a potable water source, a well or municipal water main, to the interior of a building. Once inside the building, the water-service pipe becomes a water-distribution pipe. This distinction is important, especially in zone three. When working in zone three, water-service pipe may not extend more than 5 feet (ft) into a building, unless it is of a material approved for water distribution. For example, a PE water service would have to be converted to an approved water-distribution material (Fig. 3.1) once it was in the building. But, a PB water service would not have to be converted because PB is an approved water-distribution pipe.

A water-service pipe must be rated for a pressure compatible with the pressure produced from the water source. Typically, a water service should be rated for 160 pounds per square inch (psi) at a temperature of 73.4°F. This rating will not always be applicable. If the

Figure 3.1 Plastic insert adapter in PE pipe with a copper female adapter attached.

pressure present at the water source is higher than 160 psi, the pipe's rating must be higher. On the other hand, if the water source is a well, and the water pressure from the well is below 160 psi, the rating of the pipe may be allowed to be lower. It is a good work principal to use pipe with a minimum working pressure of 160 psi even if a lower rating is approved. This allows for future changes that may increase the water pressure. The pipe must be rated with a pressure that is capable of withstanding the highest pressure developed from the water source.

Water-Service Materials

Now that you know what water service is, we are going to look at the materials approved for use for water service. The materials will be addressed in alphabetical order. Not all of the materials listed will be commonly used, but they are all approved materials. Remember, plumbing codes change and local jurisdictions have the authority to alter the code to meet local criteria. Before using these materials, or any other information in this book, consult your local code enforcement office to confirm the present status of your local code requirements.

Acrylonitrile butadiene styrene (ABS)

ABS pipe is a plastic pipe. It is normally used as a pipe for drains and vents, but if properly rated, it can be used for water service. It must meet certain specifications for pressure-rated potable water use and be approved by the local authorities. Since ABS is almost never used as a water-service pipe, it will be difficult to locate this material in a rating approved for water-service applications.

Asbestos cement pipe

Asbestos cement pipe has been used for municipal water mains in the past, but it is not used much today. It is, however, still an approved material.

Brass pipe

Brass pipe is an approved material for water-service piping, but it is rarely used. The complications of placing this metallic, threaded pipe below ground discourage its use. Brass pipe can also be used as a water-distribution pipe.

Cast-iron pipe

Cast-iron water pipe is approved for use for water service, but it would not be used for individual water supplies. Sometimes called ductile pipe, this pipe would be used in large water mains.

Copper pipe

Copper tubing is often referred to as copper pipe, but there is a difference. Copper pipe can be found with or without threads. Copper pipe is marked with a gray color code. Copper pipe is approved for water-service use, but copper tubing is used more often.

Copper tubing

Copper tubing is the copper most plumbers use. It can be purchased as soft copper, in rolls, or as rigid copper, in lengths resembling pipe. Copper tubing, frequently called copper pipe in the trade, has long been used as the plumber's workhorse. It is approved for water-service use and comes in many different grades.

The three types of copper normally used are type M, type L, and type K. The type rating refers to the wall thickness of the copper. Type K is the thickest, and type M is the thinnest, with type L in the middle. Type L is generally considered the most logical choice for water service. It offers more thickness than type M and is less expensive than type K. All three types are approved for water-service applications.

For a water-service pipe, soft, rolled copper is normally used. By using the coiled copper, plumbers can normally roll the tubing into the trench in a solid length, without joints. This reduces the risk of leaks at a later date. Copper water services are not as common as they once were. PE and PB are quickly reducing the use of copper. The reason is twofold. The plastic pipes are less expensive and are generally less affected by corrosion or other soil-related problems. Copper can also be used as a water-distribution pipe.

Chlorinated polyvinyl chloride

CPVC is a white or cream-colored plastic pipe allowed for use in water services when it is rated for potable water service. This pipe is not commonly used for water service, but it could be. CPVC is fairly fragile, especially when cold, and it requires joints if the run of pipe exceeds 20 ft. It is advisable to avoid joints in underground water services. CPVC can also be used as a water-distribution pipe.

Galvanized steel pipe

Galvanized steel pipe is an approved material for water services, but it is not a good choice. The pipe is joined together with threaded fittings. Over time, this pipe will rust. The rust can occur at the threaded areas, where the pipe walls are weakened, or inside the pipe. When the threaded areas rust, they can leak. When the interior of the pipe rusts, it can restrict the flow of water and reduce water pressure. While this gray, metal pipe is still available, it is rarely used in modern applications. If desired, galvanized pipe can be used as a water-distribution pipe.

Polybutylene

PB may very well be the pipe of the future. It is an amazing product. It is available in rolled coils and in straight lengths. PB received mixed reviews when it was introduced to the plumbing trade, but it is gaining ground fast. The problems experienced in the early years were not with the pipe but with the connections made on the pipe. These problems have been corrected, and PB is showing up in more plumbing trucks every day.

PB is approved as a water-service pipe. It is available as a water-service-only pipe and as a water-service–water-distribution pipe. If the pipe is blue in color, it is intended only for water-service use. If the pipe is gray, it can be used for water service or water distribution. Before PB pipe can be used for potable water, it must be tested by a recognized testing agency and approved by local authorities. PB is relatively inexpensive, very flexible, and a good choice for most water-service applications.

Polyethylene

PE is a black, or sometimes bluish, plastic pipe that is frequently used for water services. It resists chemical reactions, as does PB, and it is fairly flexible. This pipe is available in long coils, allowing it to be rolled out for great distances, without joints. PB may be one of the most common materials used for water services. It, however, is not rated as a water-distribution pipe. PE pipe is subject to crimping in tight turns, but it is a good pipe that will give years of satisfactory service.

Polyvinyl chloride pipe

PVC pipe is well known as a drain and vent pipe, but the PVC used for water services is not the same pipe. Both pipes are white, but the PVC used for water services must be rated for use with potable water. PVC water pipe is not acceptable as a water-distribution pipe. It is not approved for hot-water usage. Remember, when we talk of water-distribution pipes, it is assumed the building is supplied with hot and cold water. If the building's only water distribution is cold water, some of the pipes, like PVC and PE, can be used.

Water-Distribution Pipe

What is water-distribution pipe? Water-distribution pipe is the piping located inside a building that delivers potable water to plumbing fixtures. The water-distribution system normally comprises both hot and cold water. Because of this fact, the materials approved for water-distribution systems are more limited than those allowed for water-service piping.

A determining factor in choosing a pipe for water distribution is the pipe's ability to handle hot water. A water-distribution pipe must be approved for conveying hot water. In zone two, this means a pipe rated for a minimum working pressure of 100 psi, at a temperature of 180°F. Zone three requires the rating to be 80 psi, at 180°F. The reason the pressure rating is lower than that of a water-service pipe is simple. If the water pressure coming into a building exceeds 80 psi, a pressure-reducing valve must be installed at the water service to reduce the pressure to no more than 80 psi.

Water-Distribution Materials

Many water-distribution materials are also approved for use for water service. However, the reverse is not true; not all water-service materials are acceptable as water-distribution materials. Again in alphabetical order, you will find below the types of piping approved for water-distribution pipes.

Brass pipe

Brass pipe is suitable for water distribution, but it is not normally used in modern applications. While once popular, brass pipe has been replaced, in preference, by many new types of materials. The newer materials are easier to work with and usually provide longer service, with less problems.

Copper pipe and copper tubing

Both copper pipe and copper tubing are acceptable choices for water distribution. Copper tubing, sometimes mistakenly called pipe, is by far the more common choice. Copper pipe has been around for many years and has proved itself to be a good water-distribution pipe. If water has an unusually high acidic content, copper can be subject to corrosion and pinhole leaks. If acidic water is suspected, a thick-wall copper or a plastic-type pipe should be considered over the use of a thin-wall copper.

Zones two and three allow types M, L, and K to be used above and below ground, but zone one does not allow the use of type M copper when the copper is installed underground, within a building.

Galvanized steel pipe

Galvanized steel pipe remains in the approved category, but it is hardly ever used in new plumbing systems. The characteristics of galvanized pipe remove it from the competition. It is difficult to work with and is subject to rust-related problems. The rust can cause leaks and reduced water pressure and volume.

Polybutylene

PB is edging copper out of the picture in many locations. With its ease of installation, resistance to splitting during freezing conditions, and its low cost, PB is a strong competitor. Add to this the fact that PB resists chemical reactions and you have yet another advantage over copper. Many old-school plumbers are dubious of the gray, plastic pipe, but it is making a good reputation for itself.

Drain Waste and Vent Pipe

Pipe used for the DWV system is considerably different from its water pipe cousins. The most noticeable difference is size. DWV pipes typically range in size from 1½ to 4 in, in diameter; some are smaller and some are much larger. When first glancing at the names of DWV pipe materials, they may seem the same as the water pipe variety, but there are differences. Remember, for plastic water pipes to carry potable water, they must be tested and approved for potable water use.

There are a number of materials approved for DWV purposes, but in practice, only a few are commonly used in modern plumbing applications. Let's look, in alphabetical order, at the materials approved for DWV systems.

Acrylonitrile butadiene styrene

ABS pipe is black, or sometimes a dark gray color, and will be labeled as a DWV pipe when it is meant for DWV purposes. ABS is normally used as a DWV pipe instead of as a water pipe. The standard weight rating for common DWV pipe is schedule 40.

ABS pipe is easy to work with and may be used above or below ground. It cuts well with a hack saw or regular handsaw. This material is joined with a solvent-weld cement and rarely leaks, even in less than desirable installation circumstances. ABS was very popular for a long time, but in many areas it is being pushed aside by PVC pipe, the white plastic DWV pipe. ABS is extremely durable and can take hard abuse without breaking or cracking.

In zone one, the use of ABS is restricted to certain types of structures. ABS may not be used in buildings that have more than three habitable floors. The building may have a buried basement, where at least one-half of the exterior wall sections are at ground level or below. The basement may not be used as habitable space. So, it is possible to use ABS in a four-story building so long as the first story is buried in the ground, as stated above, and not used as living space.

Aluminum tubing

Aluminum tubing is approved for above-ground use only. Aluminum tubing may not be allowed in zone one. Aluminum tubing is usually joined with mechanical joints and coated to prevent corrosive action. This material is, like most others, available in many sizes. The use of aluminum tubing has not become common for average plumbing installations in most regions.

Borosilicate glass

Borosilicate glass pipe may be used above or below ground for DWV purposes in zone two. Underground use requires a heavy schedule of pipe. Zones one and three do not recognize this pipe as an approved material. However, as with all regulations, local authorities have the power to amend regulations to suit local requirements.

Brass pipe

Brass pipe could be used as a DWV pipe, but it rarely is. The degree of difficulty in working with it is one reason it is not used more often. Zones two and three don't allow brass pipe to be used below grade for DWV purposes.

Cast-iron pipe

Cast-iron pipe has long been a favored DWV pipe. Cast iron has been used for many years and provides good service for extended periods of time. The pipe is available in a hub-and-spigot style, the type used years ago, and in a hubless version. The hubless version is newer and is joined with mechanical joints, resembling a rubber coupling (Fig. 3.2), surrounded and compressed by a stainless steel band.

The older, bell-and-spigot, or hub-and-spigot, type of cast iron is what is most often encountered during remodeling jobs. This type of cast iron was normally joined with oakum and molten lead. Today, however, there are rubber adapters available for creating joints with this type of pipe. These rubber adapters will also allow plastic pipe to be mated to the cast iron.

Cast-iron pipe is frequently referred to as soil pipe. This nickname separates DWV cast iron from cast iron designed for use as a potable water pipe. Cast iron is available as a service-weight pipe and as an extra-heavy pipe. Service-weight cast iron is the most commonly used. Even though the cost of labor and material for installing cast iron is more than it is for schedule 40 plastic, cast iron still sees frequent use, both above and below grade.

Cast iron is sometimes used in multifamily dwellings and custom homes to deaden the sound of drainage as it passes down the pipe in walls adjacent to living space. If chemical or heat concerns are present, cast iron is often chosen over plastic pipe.

Copper

Copper pipe is made in a DWV rating. This pipe is thin walled and identified with a yellow marking. The pipe is a good DWV pipe, but it

Figure 3.2 Hubless band for Cast-iron connections.

is expensive and time-consuming to install. DWV copper is not normally used in new installations unless extreme temperatures, such as those from a commercial dishwasher, warrant the use of a nonplastic pipe. DWV copper is approved for use above and below ground in zones one and two. Zone three requires a minimum copper rating of type L for copper used underground for DWV purposes.

Galvanized steel pipe

Galvanized steel pipe keeps popping up as an approved material, but it is no longer a good choice for most plumbing jobs. As galvanized pipe ages and rusts, the rough surface, from the rust, is prone to catching debris and creating pipe blockages. Another disadvantage to galvanized DWV pipe is the time it takes to install the material.

Galvanized pipe is not allowed for underground use in DWV systems. When used for DWV purposes, galvanized pipe should not be installed closer than 6 in to the earth.

Lead pipe

Lead pipe is still an approved material, but like galvanized pipe, it has little place in modern plumbing applications. Zone two does not allow the use of lead for DWV installations. Zone three limits the use of lead to above-grade installations.

Polyvinyl chloride pipe

PVC is probably the leader in today's DWV pipe. This plastic pipe is white and is normally used in a rating of schedule 40. PVC pipe uses a solvent weld joint and should be cleaned and primed before being glued together. This pipe will become brittle in cold weather. If PVC is dropped on a hard surface while the pipe is cold, it is likely to crack or shatter. The cracks can go unnoticed until the pipe is installed and tested. Finding a cracked pipe moments before an inspector is to arrive is no fun, so be advised: handle cold PVC with care. PVC may be used above or below ground.

In zone one, the use of PVC is restricted to certain types of structures. PVC may not be used in buildings that have more than three habitable floors. The building may have a buried basement, where at least one-half of the exterior wall sections are at ground level or below. The basement may not be used as habitable space. So, it is possible to use PVC in a four-story building so long as the first story is buried in the ground, as stated above, and not used as living space.

If the underground piping will be used as a building sewer, one of the following types of pipes may be used:

- ABS
- Cast iron
- Vitrified clay
- PVC
- Concrete
- Asbestos cement

In zone three, bituminized-fiber pipe, type L copper pipe, and type K copper pipe may be used. Zone one tends to stick to the general guidelines, as given in the previous paragraphs.

If a building sewer will be installed in the same trench that contains a water service, some of the above pipes may not be used. Standard procedure for pipe selection under these conditions calls for the use of a pipe approved for use inside a building. These types of pipes could include ABS, PVC, and cast iron. Pipes that are more prone to breakage, like a clay pipe, are not allowed unless special installation precautions are taken.

Sewers installed in unstable ground are also subject to modified rulings. Normally, any pipe approved for use underground, inside a building, will be approved for use with unstable ground. But, the pipe must be well supported for its entire length.

Chemical wastes must be conveyed and vented with a system separated from the building's normal DWV system. The material requirements for chemical-waste piping must be obtained from the local code enforcement office.

Storm Drainage Materials

The materials used for interior and underground storm drainage may generally be the same materials used for sanitary drainage. Any approved DWV material is normally approved for use in storm drainage.

Inside storm drainage

Materials commonly approved in zone three for interior storm drainage include the following:

- ABS
- Type DWV copper
- Type L copper
- Asbestos cement
- Cast iron
- PVC
- Type M copper
- Type K copper
- Bituminized fiber
- Concrete

- Vitrified clay
- Brass
- Galvanized

- Aluminum
- Lead

Zone two's approved pipe materials for interior storm drainage include:

- ABS
- Type DWV copper
- Type L copper
- Asbestos cement
- Concrete
- Galvanized
- Brass

- PVC
- Type M copper
- Type K copper
- Cast iron
- Aluminum
- Black steel
- Lead

Zone one follows its standard pipe approvals for storm-water piping. Storm-drainage sewers are a little different. If you are installing a storm-drainage sewer, use one of the following types of pipes:

Zone one. Follow the basic guidelines for approved piping.

Zone two

- Cast iron
- Vitrified clay
- ABS
- Aluminum (coated to prevent corrosion)

- Asbestos cement
- Concrete
- PVC

Zone three

- Cast iron
- Vitrified clay
- ABS
- Bituminized fiber
- Type K copper
- Type DWV copper

- Asbestos cement
- Concrete
- PVC
- Type L copper
- Type M copper

Subsoil Drains

Subsoil drains are designed to collect and drain water entering the soil. They are frequently slotted pipes and could be made from any of the following materials:

- Asbestos cement
- Vitrified clay
- Cast iron
- Bituminized fiber
- PVC
- PE

Zone two does not allow bituminized fiber pipe to be used as a sub-soil drain. It may not allow cast iron or some plastics. Zone one sticks to its normal pipe approvals. As always, check with local authorities before using any material.

Other Types of Materials

We have concluded our look at the various types of pipes approved for plumbing, but there are other types of materials to take into consideration. Valves, fittings, and nipples all fall under the watchful eye of the code enforcement office. In addition to these materials, there are still others to be discussed.

Fittings

Fittings that are made from cast iron, copper, plastic, steel, and other types of iron are all approved for use in their proper place. Generally speaking, fittings must either be made from the same material as the pipe they are being used with or be compatible with the pipe.

Valves

Valves have to meet some standards, but most of the decision for the use of valves will come from the local code enforcement office. Valves, like fittings, must be either of the same material as the pipe they are being used with or compatible with the pipe. Size and construction requirements will be stipulated by local jurisdictions.

Nipples

Manufactured pipe nipples are normally made from brass or steel. These nipples range in length from ⅛ to 12 in. Nipples must live up to certain standards, but they should be rated and approved before you are able to obtain them.

Flanges

Closet flanges (Fig. 3.3) made from plastic must have a thickness of ¼ in. Brass flanges may have a thickness of only ⅛ in. Flanges intended for caulking must have a thickness of ¼ in, with a caulking depth of 2 in. The screws or bolts used to secure flanges to a floor must be brass. All flanges must be approved for use by the local authorities.

Figure 3.3 Closet flange.

Zone two prohibits the use of offset flanges (Fig. 3.4), without prior approval. Zone three requires hard-lead flanges to weigh at least 25 ounces (oz) and to be made from a lead ally with no less than a 7.75 percent antimony, by weight. Zone one requires flanges to have a diameter of about 7 in. In zone one, the combination of the flange and the pipe receiving it must provide about 1½ in of space to accept the wax ring or sealing gasket.

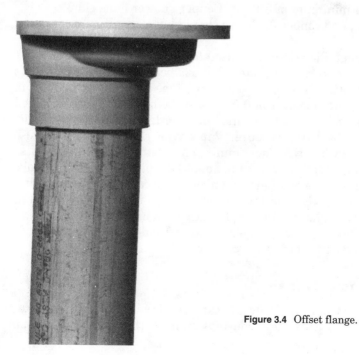

Figure 3.4 Offset flange.

Cleanout plugs

Cleanout plugs will be made of plastic or brass. Brass plugs are to be used only with metallic fittings. Unless they create a hazard, cleanout plugs shall have raised, square heads. If located where a hazard from the raised head may exist, countersunk heads may be used. Zone two requires borosilicate glass plugs to be used with cleanouts installed on borosilicate pipe.

Fixtures

Plumbing fixtures are regulated and must have smooth surfaces. These surfaces must be impervious. All fixtures must be in good working order and may not contain any hidden surfaces that may foul or become contaminated.

The rest of them

Lead bends and traps are not used much anymore, but if you decide to use these items, check with your code officer for guidance. These units must have a wall thickness of at least ⅛ in.

The days of using sheet lead and copper to form shower pans are all but gone, but the code does still offer minimum standards for these materials. Lead shower pans must not be rated at less than 4 pounds per square foot (lb/ft²). If you need to use a lead pipe flashing, it should be rated at a minimum of 3 lb/ft². Copper shower pans should weigh in at 12 oz/ft², and copper flashings should have a minimum weight of 8 oz/ft².

When nonmetallic material is used for a shower pan, it must meet minimum standards. The material must be marked to indicate its approved qualities. Normally, membrane-type material is required to have a minimum thickness of 0.040 in. If the material must be joined together at seams, it must be joined in accordance with the manufacturer's recommended procedure. Paper-type shower pans are also allowed for use when they meet minimum construction requirements.

Soldering bushings, once used to adapt lead pipe to other materials, and caulking ferrules, used in the conversion of cast iron to other materials, are all but a thing of the past. Lead pipe is usually removed in today's plumbing and rubber adapters are used in place of caulking ferrules. If you have a nostalgic interest in these old-school items, you can find standards for them in your code book.

Connecting Your Materials

You now know what materials you can use for various purposes; next, you are going to learn how you may connect them. The connection methods approved are normally approved on a local level. But, there are some basics, and that's what follows.

Compatibility and performance

The main considerations in a good connection are compatibility and performance. A connection must be able to endure the pressure exerted on it in a normal testing procedure for the plumbing. For DWV connections, this usually amounts to a pressure of between 4 and 5 psi. On water pipes, the pressure for a test should be equal to the highest pressure expected to be placed on the system once it is in use. Zone two requires a pressure test with a pressure of at least 25 psi more than the highest working pressure.

Basic preparation

Before pipe is connected with fittings, the pipe should be properly prepared. This means cutting the pipe evenly and clearing it of any burrs or obstructions. If a connection will be made with a material of different construction than that of the pipe, the connector must be compatible with the pipe. For example, a rubber coupling, held in place with stainless steel clamps, can be used to join numerous types of different pipe styles.

ABS and PVC

Plastic pipes should be joined with solvent cements designed for the specific type of pipe being worked with. The plastic pipe and fittings should be clean, dry, and grease-free before a joint is made. When working with plastic pipes, a cleaner and primer are often recommended, prior to the application of solvent cement. Always follow the manufacturer's recommendations for joining pipe and fittings. For best results, once the pipe has been coated with cement and inserted in a fitting, it should be turned about a quarter of a turn. This helps to spread the glue and make a better joint.

ABS pipe and fittings tend to harden much quicker than PVC. ABS is also less sensitive to dirt and water in its joints. This is not to say that you can ignore proper procedures with ABS, but it is more forgiving than PVC.

Unusual pipes

Some types of materials, like asbestos cement pipe and bituminized-fiber pipe, only lend themselves to mechanical joints. Refer to manufacturer's recommendations, your code book, and the local code officer for the best methods in joining this type of pipe.

Cast-iron pipe

Cast-iron pipe offers a myriad of ways to join it with fittings. You can use hot lead, rubber doughnuts, rubber couplings, or special bands designed for use with hubless cast iron, just to name a few.

In the old days, cast iron was almost always joined with a caulked joint, using molten lead. This procedure is still used today. To make this kind of joint, oakum or hemp is placed into the hub of a cast-iron fitting, around the pipe being joined to the fitting. The oakum or hemp must be dry when it is installed if it is to make a good joint. Once the packing is in place, molten lead is poured into the hub. Once the lead is poured, a packing tool, basically a special chisel, is used with a hammer to drive the lead down deeper into the hub. Once the oakum becomes wet, it expands and seals the joint.

Without experience or an instructor, this type of joint can be extremely dangerous to make. The hot lead can take the skin right off your bones, and if the hot lead comes into contact with a wet surface, it can explode, causing personal injury. Proper clothing and safety gear are a necessity on this type of job, even for seasoned professionals.

If you are using lead to make joints in cast iron meant to carry potable water, you will not use oakum or hemp. Instead, you will use a rope packing. This type of packing has a higher density than oakum and is meant for use with potable water installations.

Rubber doughnuts, as they are called in the trade, offer an alternative to caulking with hot lead. These special rubber adapters are placed in the hub of a fitting and lubricated. The end of the pipe to be joined is inserted into the rubber gasket and driven or pulled into place. There are special tools used to join soil pipe in this manner (Fig. 3.5), but some plumbers use a block of wood and a sledge hammer to drive the pipe into the doughnut.

If a mechanical joint is being used on cast iron meant for potable water, there must be an elastomeric gasket on the joint, held in place with an approved flange. One example of a mechanical joint with an

Figure 3.5 Soil pipe assembly tool.

elastomeric gasket and approved flange is the standard band used with hubless cast iron, but there are other types of mechanical joints available and approved for these types of joints.

Copper

When it comes to copper, your options for joints include compressions fittings (Fig. 3.6), soldered joints, screw joints, and other types of mechanical joints. Solder used to join copper for potable water systems must contain less than 0.02 percent lead. Threaded joints must be sealed with an acceptable pipe compound or tape. Welded and brazed joints offer two other ways to join your copper. Flared joints are still another way to mate copper to its fittings (Fig. 3.7). Unions (Figs. 3.8 and 3.9) are allowed for connecting copper and are usually required with the installation of water heaters.

CPVC pipe

CPVC pipe is the homeowner's friend. Homeowners flock to this plastic water pipe because all they have to do is glue it together, or at least, that's what they think. CPVC is a finicky pipe. It must be primed with

Figure 3.6 Compression tee.

Figure 3.7 Flare joint.

Figure 3.8 Closed union.

Figure 3.9 Open union.

an approved primer before it is glued. If this step is ignored, the joint will not be as strong as it should be. CPVC also requires a long time for its joints to set up. This pipe must be clean, dry, and grease-free before you begin the connection process. Don't eliminate the priming process; it is essential for a good joint.

Galvanized pipe

Galvanized pipe lends itself to threaded connections, but rubber couplings can also be used. Use the large rubber couplings, not the hubless cast-iron type (Fig. 3.10). Remember to apply the appropriate pipe dope or tape to the threads if you are making a screw connection. There are other types of approved ways to connect galvanized pipe, but these are the easiest.

Figure 3.10 Rubber coupling.

Polybutylene

PB pipe is normally connected to insert fittings with special clamps (Fig. 3.11). These clamps are installed with a crimping tool designed for use with PB pipe (Fig. 3.12). Compression fittings are also allowed with PB pipe, but the ferrules should be nonmetallic to avoid cuts in the pipe. Flaring is possible with a special tool, but it is rarely needed. Another method for joining PB pipe is heat fusion. This process is

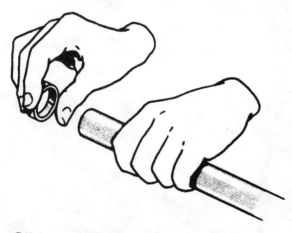

Slide a copper crimp ring over the pipe end.

Figure 3.11 PB crimp ring. (*Courtesy of Vanguard Plastics, Inc.*)

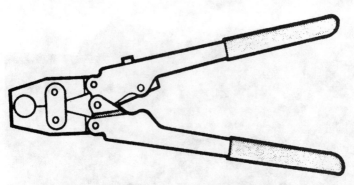

Figure 3.12 PB crimp tool. (*Courtesy of Vanguard Plastics, Inc.*)

not normally used with water-distribution piping; crimp rings are the most common method of joining PB pipe in a water-distribution system.

Polyethylene

PE pipe is typically joined with insert fittings and stainless steel clamps (Fig. 3.13). The insert fitting is placed inside the pipe and a stainless steel clamp is applied outside of it over the insert fitting's shank. For best results, use two clamps.

Figure 3.13 Plastic insert tee in PE pipe.

Fixtures

Fixtures must be connected to their drains in an approved manner. For sinks and such, the typical connection is made with slip nuts and washers. The washers may be rubber or nylon. For toilets, the seal is normally made with a wax ring.

Pipe penetrations

Pipe penetrations must be sealed to protect against water infiltration, the spreading of fire, and rodent activity. When a pipe penetrates a wall or roof, it must be adequately sealed to prevent the above-mentioned problems. Many styles of roof flashings are available for sealing pipe holes (Figs. 3.14 and 3.15).

Reminder Notes

Zone one

1. Type M copper tubing may not be used below grade when used inside a building.
2. ABS and PVC pipe for drainage is restricted to use in buildings having no more than three levels of habitable space. A fourth level is

Figure 3.14 Metal roof flange.

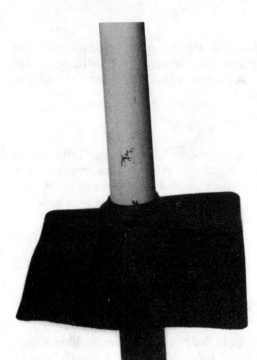

Figure 3.15 Flexible roof flange.

allowed under certain conditions. The conditions are that the first of the four levels must have no more than one-half of its exterior wall surface above grade, and this level may not be used as living space.

3. Aluminum tubing may not be approved for use as a DWV pipe. Check with local regulations.

4. Closet flanges should have a diameter of about 7 in. When combined with the pipe receiving it, the flange should offer an area of about 1½ in to accept a wax ring or sealing gasket.

Zone two

1. Pipe used to convey hot water must have a minimum working pressure of 100 psi, at 180°F.

2. Brass pipe may not be used for DWV purposes below ground level.

3. Borosilicate glass pipe is approved for above and below ground use in DWV systems.

4. Lead pipe may not be used in a new DWV system.

5. Copper pipe and bituminized-fiber pipe are not allowed for use as building sewers.

6. Zone two allows coated aluminum tubing to be used for underground storm-water sewer applications.

7. Zone two does not allow bituminized-fiber pipe for subsoil drains. It may not allow cast-iron pipe and some plastic pipes.

8. Offset closet flanges are prohibited without prior approval.

9. Borosilicate-glass cleanouts are required to have borosilicate glass plugs.

10. When testing water pipe, the test pressure must be at least 25 psi higher than the highest working pressure.

Zone three

1. Water-service pipe must not extend more than 5 ft inside a building before being converted to a water-distribution pipe unless the pipe used for the water service is also approved for water distribution.

2. Pipe used to convey hot water must have a minimum working pressure of 80 psi, at 180°F.

3. Brass pipe may not be used for DWV purposes below ground level.

4. When used for underground DWV purposes, copper must have a minimum rating of type L.

5. Lead pipe, when used as a DWV material, is limited to use above ground.

6. Zone three allows the use of vitrified-clay and bituminized-fiber pipe in interior storm-water applications.

7. Zone three allows the use of vitrified-clay, bituminized-fiber, and copper pipe in storm-water sewer applications.

8. Lead closet flanges must weigh at least 25 oz and be made from a lead alloy with no less than 7.75 percent antimony, by weight.

4

Drainage Systems

Drainage systems intimidate many people. When these people look at their code books, they see charts and math requirements that make them nervous. Their fear is largely unjustified. For the inexperienced, the fundamentals of building a suitable drainage system can appear formidable. But, with a basic understanding of plumbing, the process becomes much less complicated. This chapter is going to take you, step by step, through the procedures of making a working drainage system.

You are going to learn the criteria for sizing pipe. You will be shown which types of fittings can be used in various applications. During the process, you will be given instructions for the proper installation of a drainage system. Then, the focus will shift to indirect and special wastes. Storm drainage will complete the chapter. By the end of this chapter, you will be quite knowledgeable on the subject of drainage systems.

Pipe Sizing

Sizing pipe for a drainage system is not difficult. You must know a few benchmark numbers, but you don't have to memorize these numbers. Your code book will have charts and tables that provide the benchmarks. All you must know is how to interpret and use the information provided.

The size of a drainage pipe is determined by using various factors, the first of which is the drainage load. This refers to the volume of drainage the pipe will be responsible for carrying. When you refer to your code book, you will find ratings that assign a fixture-unit value to various plumbing fixtures. For example, a residential toilet has a fixture-unit value of 4. A bathtub's fixture-unit value is 2.

By using the ratings given in your code book, you can quickly assess the drainage load for the system you are designing. Since plumbing fixtures require traps, you must also determine what size traps are required for particular fixtures. Again, you don't need a math degree to accomplish this task. In fact, your code book will tell you what trap sizes are required for most common plumbing fixtures.

Your code book will provide trap-size requirements for specific fixtures. For example, by referring to the ratings in your code book, you will find that a bathtub requires a 1½-in trap. A lavatory may be trapped with a 1¼-in trap. The list will go on to describe the trap needs for all common plumbing fixtures. Trap sizes will not be provided for toilets, since toilets have integral traps.

When necessary, you can determine a fixture's drainage-unit value by the size of the fixture's trap. A 1¼-in trap, the smallest trap allowed, will carry a fixture-unit rating of 1. A 1½-in trap will have a fixture-unit rating of 2. A 2-in trap will have a rating of 3 fixture units. A 3-in trap will have a fixture-unit rating of 5, and a 4-in trap will have a fixture-unit rating of 6. This information can be found in your code book and may be applied for a fixture not specifically listed with a rating in your code book. Table 4.1 shows this information for zone two. Table 4.2 shows the same information for zone three. Zone one uses a different rating for the fixture-unit loads on traps. Refer to Table 4.3 for zone one's requirements.

Determining the fixture-unit value of a pump does require a little math, but it's simple. By taking the flow rate, in gallons per minute (gpm), assign 2 fixture units for every gpm of flow. For example, a pump with a flow rate of 30 gpm would have a fixture-unit rating of 60. Zone three is more generous. In zone three, 1 fixture unit is assigned for every 7½ gpm. With the same pump, producing 30 gpm, zone three's

TABLE 4.1   Zone Two's Fixture-Unit Requirements on Trap Sizes

Trap size (in)	No. of fixture units
1¼	1
1½	2
2	3
3	5
4	6

TABLE 4.2   Zone Three's Fixture-Unit Requirements on Trap Sizes

Trap size (in)	No. of fixture units
1¼	1
1½	2
2	3
3	5
4	6

TABLE 4.3   Zone One's Fixture-Unit Requirements on Trap Sizes

Trap size	No. of fixture units
1¼	1
1½	3
2	4
3	6
4	8

fixture-unit rating would be 4. That's quite a difference from the ratings in zones one and two.

Other considerations when sizing drainage pipe is the type of drain you are sizing and the amount of fall that the pipe will have. For example, the sizing for a sewer will be done a little differently than the sizing for a vertical stack. A pipe with a ¼-in fall will be rated differently than the same pipe with a ⅛-in fall.

Sizing Building Drains and Sewers

Building drains and sewers use the same criteria in determining the proper pipe size. The two components you must know to size these types of pipes are the total number of drainage fixture units entering the pipe and the amount of fall placed on the pipe. The amount of fall is based on how much the pipe drops in each foot it travels. A normal grade is generally ¼ in/ft, but the fall could be more or less.

When you refer to your code book you will find information, probably a table, to aid you is sizing building drains and sewers. Let's take a look at how a building drain for a typical house would be sized in zone three.

Sizing—Example 1

Our sample house has two and one-half bathrooms, a kitchen, and a laundry room. To size the building drain for this house, we must determine the total fixture-unit load that may be placed on the building drain. To do this, we start by listing all of the plumbing fixtures producing a drainage load. In this house we have the following fixtures:

- One bathtub
- Three toilets
- One shower
- Three lavatories

- One kitchen sink
- One dishwasher
- One laundry tub
- One clothes washer

By using Table 4.4, we can determine the number of drainage fixture units assigned to each of these fixtures. When we add up all the fixture units, we have a total load of 28 fixture units. It is always best to allow

TABLE 4.4   Fixture-Unit Ratings in Zone Three

Fixture	Rating
Bathtub	2
Shower	2
Residential toilet	4
Lavatory	1
Kitchen sink	2
Dishwasher	2
Clothes washer	3
Laundry tub	2

a little extra in your fixture-unit load so your pipe will be in no danger of becoming overloaded. The next step is to look at Table 4.5 to determine the sizing of our building drain. Table 4.6 shows pitch allowances for zone two, and Table 4.7 shows standards for zone one.

Our building drain will be installed with a ¼-in fall. By looking at Table 4.8, we see that we can use a 3-in pipe for our building drain, based on the number of fixture units, but, notice the footnote below the chart. The note indicates that a 3-in pipe may not carry the discharge of more than two toilets, and our test house has three toilets. This means we will have to move up to a 4-in pipe.

Suppose our test house only had two toilets, what would the outcome be then? If we eliminate one of the toilets, our fixture load drops to 24. According to the table, we could use a 2½-in pipe, but we know our

TABLE 4.5   Zone Three's Minimum Drainage-Pipe Pitch

Pipe diameter (in)	Pitch (in/ft)
Under 3	¼
3–6	⅛
8 or larger	⅟₁₆

TABLE 4.6   Zone Two's Minimum Drainage-Pipe Pitch

Pipe diameter (in)	Pitch (in/ft)
Under 3	¼
3 or larger	⅛

TABLE 4.7   Zone One's Minimum Drainage-Pipe Pitch

Pipe diameter (in)	Pitch (in/ft)
Under 4	¼
4 or larger	⅛

TABLE 4.8   Building-Drain Sizing Table for Zone Three

Pipe size (in)	Pipe grade (in/ft)	Maximum no. of fixture units
2	¼	21
3	¼	42*
4	¼	216

*No more than two toilets may be installed on a 3-in building drain.

building drain must be at least a 3-inch pipe, to connect to the toilets. A fixture's drain may enter a pipe the same size as the fixture drain or a pipe that is larger, but it may never be reduced to a smaller size, except with a 4- by 3-in closet bend.

So, with two toilets, our sample house could have a building drain and sewer with a 3-in diameter. But, should we run a 3- or a 4-in pipe? In a highly competitive bidding situation, 3-in pipe would probably win the coin toss. It would be less expensive to install a 3-in drain, and you would be more likely to win the bid on the job. However, when feasible, it would be better to use a 4-in drain. This allows the homeowner to add another toilet at some time in the future. If you install a 3-in sewer, the homeowner would not be able to add a toilet without replacing the sewer with 4-in pipe.

Horizontal Branches

Horizontal branches are the pipes that branch off from a stack to accept the discharge from fixture drains. These horizontal branches normally leave the stack as a horizontal pipe, but they may turn to a vertical position, while retaining the name of horizontal branch. The procedure for sizing a horizontal branch is similar to the one used to size a building drain or sewer, but the ratings are different. Your code book will contain the benchmarks for your sizing efforts, but here are some examples.

The number of fixture units allowed on a horizontal branch is determined by pipe size and pitch. All of the following examples are based on a pitch of 1¼ in/ft. A 2-in pipe can accommodate up to 6 fixture units, except in zone one, where it can have 8 fixture units. A 3-in pipe can handle 20 fixture units but not more than two toilets. In zone one, a 3-in pipe is allowed up to 35 units and up to three toilets. A 1½-in pipe will carry 3 fixture units, unless you are in zone one. Zone one only allows a 1½-in pipe to carry 2 fixture units, and they may not be from sinks, dishwashers, or urinals. A 4-in pipe will take up to 160 fixture units, except in zone one, where it will take up to 216 units. Table 4.9 gives you an example of how a table for sizing horizontal fixture units might be assembled in your code book.

TABLE 4.9   Example of Horizontal-Branch Sizing Table in Zone Two*

Pipe size (in)	Maximum no. of fixture units on a horizontal branch
1¼	1
1½	2
2	6
3	20[†]
4	160
6	620

* Table does not represent branches of the building drain, and other restrictions apply under battery-venting conditions.
[†] Not more than two toilets may be connected to a single 3-in horizontal branch. Any branch connecting with a toilet must have a minimum diameter of 3 in.

Stack Sizing

Stack sizing is not too different from the other sizing exercises we have studied. When you size a stack, you must base your decision on the total number of fixture units carried by the stack and the amount of discharge into branch intervals. This may sound complicated, but it isn't.

Look at Tables 4.10 and 4.11 You will notice that there are three columns. The first is for pipe size, the second represents the discharge of a branch interval, and the last column shows the ratings for the total fixture-unit load on a stack. This table is based on a stack with no more than three branch intervals. See Fig. 4.1 for an example of what is meant by the limit of three branch intervals.

TABLE 4.10   Stack-Sizing Table for Zone Three

Pipe size (in)	Fixture-unit discharge on stack from a branch	Total fixture units allowed on stack
1½	2	4
2	6	10
3	20*	48*
4	90	240

* No more than two toilets may be placed on a 3-in branch, and no more than six toilets may be connected to a 3-in stack.

TABLE 4.11   Stack-Sizing Table for Zone Two

Pipe size (in)	Fixture-unit discharge on stack from a branch	Total fixture units allowed on stack
1½	3	4
2	6	10
3	20*	30*
4	160	240

* No more than two toilets may be placed on a 3-in branch, and no more than six toilets may be connected to a 3-in stack.

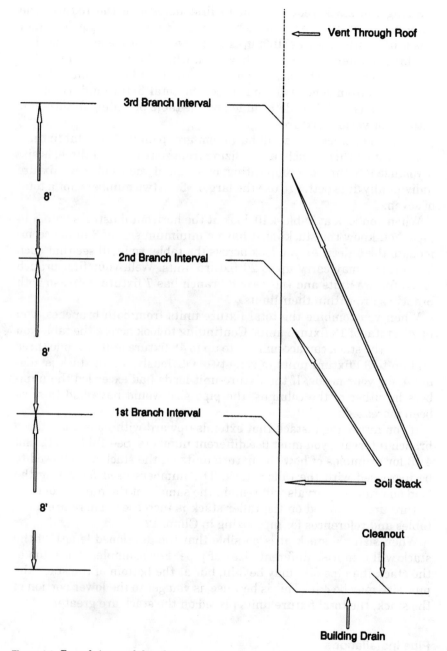

Figure 4.1 Branch-interval detail.

Sizing the stack requires you to first determine the fixture load entering the stack at each branch interval. Here is an example of how this type of sizing works. In our example we will size a stack that has two branch intervals. Fig. 4.2 shows you what the stack and branches look like. The lower branch has a half-bath and a kitchen on it. Using the ratings from zones two and three, the total fixture-unit count for this branch is 6. This is determined by the table providing fixture-unit ratings for various fixtures.

The second stack has a full bathroom group on it. The total fixture-unit count on this branch, using sizing from zone two and three, is six, if you use a bathroom group rating, or seven, if you count each fixture individually. It is better to use the larger of the two numbers, for a total of seven.

When you look at Table 4.10, look at the horizontal listings for a 3-in pipe. You know the stack must have a minimum size of 3 in to accommodate the toilets. As you look across the table, you will see that each 3-in branch may carry up to 20 fixture units. Well, your first branch has 6 fixture units and the second branch has 7 fixture units, so both branches are within their limits.

When you combine the total fixture units from both branches, you have a total of 13 fixture units. Continuing to look across the table you see that the stack can accommodate up to 48 fixture units in zone three and up to 30 fixture units in zone two. Obviously, a 3-in stack is adequate for your needs. If the fixture-unit loads had exceeded the numbers in either of the columns, the pipe size would have had to have been increased.

If you are sizing a stack that extends upward with more than three branch intervals, you must use different numbers. See Tables 4.12 and 4.13 for examples of how the fixture units on the stack are allowed to be increased with the taller stack. The numbers used for rating the load on branch intervals will remain the same, but the number of total fixture units allowed on the taller stack is increased. There are more tables and references for pipe sizing in Chap. 12.

When sizing a stack, it is possible that the developed length of the stack will comprise different sizes of pipe. For example, at the top of the stack the pipe size may be 3 in, but at the bottom of the stack the pipe size may be 4 in. This is because as you get to the lower portion of the stack, the total fixture units placed on the stack are greater.

Pipe Installations

Once the pipe is properly sized, it is ready for installation. There are a few regulations pertaining to pipe installation that you need to be aware of.

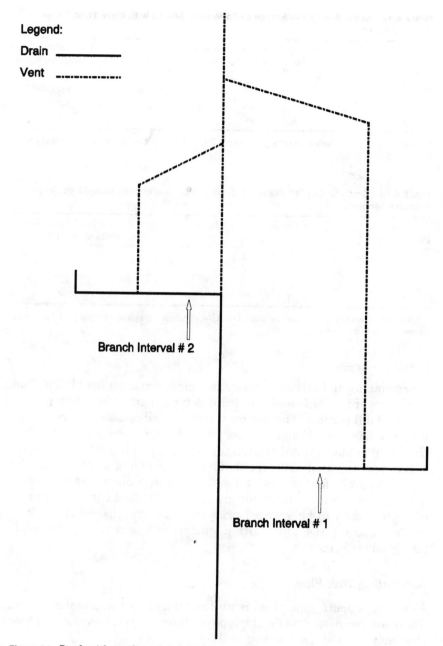

Figure 4.2 Stack with two branch intervals.

TABLE 4.12   Stack-Sizing Tall Stacks in Zone Two (Stacks with More Than Three Branch Intervals)

Pipe size (in)	Fixture-unit discharge on stack from a branch	Total fixture units allowed on stack
1½	2	8
2	6	24
3	16*	60*
4	90	500

* No more than two toilets may be placed on a 3-in branch, and no more than six toilets may be connected to a 3-in stack.

TABLE 4.13   Stack-Sizing Tall Stacks in Zone Three (Stacks with More Than Three Branch Intervals)

Pipe size (in)	Fixture-unit discharge on stack from a branch	Total fixture units allowed on stack
1½	2	8
2	6	24
3	20*	72*
4	90	500

* No more than two toilets may be placed on a 3-in branch, and no more than six toilets may be connected to a 3-in stack.

Grading your pipe

When you install horizontal drainage piping, you must install it so that it falls toward the waste-disposal site. A typical grade for drainage pipe is ¼ in of fall per foot. This means the lower end of a 20-ft piece of pipe would be 5 in lower than the upper end, when properly installed. While the ¼-in/ft grade is typical, it is not the only acceptable grade for all pipes.

If you are working with pipe that has a diameter of 2½ in, or less, the minimum grade for the pipe is ¼ in/ft. Pipes with diameters between 3 and 6 in are allowed a minimum grade of ⅛ in/ft. Zone one requires special permission to be granted prior to installing pipe with a ⅛-in/ft grade. In zone three, pipes with diameters of 8 in or more, an acceptable grade is ¹⁄₁₆ in/ft.

Supporting Your Pipe

How you support your pipes is also regulated by the plumbing code. There are requirements for the type of materials you may use and how they may be used. Let's see what they are.

One concern with the type of hangers used is their compatibility with the pipe they are supporting. You must use a hanger that will not have a detrimental effect on your piping. For example, you may not use galvanized strap hanger to support copper pipe. As a rule of thumb, the hangers used to support a pipe should be made from the same material as the pipe being supported. For example, copper pipe should be hung with copper hangers. This eliminates the risk of a corrosive action between two different types of materials. If you are using a plastic or plastic-coated hanger, you may use it with all types of pipe. The exception to this rule might be when the piping is carrying a liquid with a temperature that might affect or melt the plastic hanger.

The hangers used to support pipe must be capable of supporting the pipe at all times. The hanger must be attached to the pipe and to the member holding the hanger in a satisfactory manner. For example, it would not be acceptable to wrap a piece of wire around a pipe and then wrap the wire around the bridging between two floor joists. Hangers should be securely attached to the member supporting it. For example, a hanger should be attached to the pipe and then nailed to a floor joist. The nails used to hold a hanger in place should be made of the same material as the hanger if corrosive action is a possibility.

Both horizontal and vertical pipes require support. The intervals between supports will vary, depending upon the type of pipe being used and whether it is installed vertically or horizontally. The following examples will show you how often you must support the various types of pipes when they are hung horizontally. These examples are the maximum distances allowed between supports for zone three:

- ABS—every 4 in
- Galvanized—every 12 in
- DWV copper—every 10 in
- Cast iron—every 5 in
- PVC—every 4 in

When these same types of pipes are installed vertically, in zone three, they must be supported at no less than the following intervals:

- ABS—every 4 in
- Galvanized—every 15 in
- DWV copper—every 10 in
- Cast iron—every 15 in
- PVC—every 4 in

Table 4.14 for horizontal pipe support intervals required in zone one and Table 4.15 for vertical support intervals. For zone two, see Table 4.16 for horizontal support requirements and Table 4.17 for vertical support intervals.

TABLE 4.14   Horizontal Pipe-Support Intervals in Zone One

Support material	Maximum distance of supports (ft)
ABS	4
Cast iron	At each pipe joint*
Galvanized (1 in and larger)	12
Galvanized (¾ in and smaller)	10
PVC	4
Copper (2 in and larger)	10
Copper (1½ in and smaller)	6

* Cast-iron pipe must be supported at each joint, but supports may not be more than 10 ft apart.

TABLE 4.15   Vertical Pipe-Support Intervals in Zone One

Type of drainage pipe	Maximum distance of supports*
Lead pipe	4 ft
Cast iron	At each story
Galvanized	At least every other story
Copper	At each story†
PVC	Not mentioned
ABS	Not mentioned

* All stacks must be supported at their bases.
† Support intervals may not exceed 10 ft.

TABLE 4.16   Horizontal Pipe-Support Intervals in Zone Two

Type of drainage pipe	Maximum distance of supports (ft)
ABS	4
Cast iron	At each pipe joint
Galvanized (1 in and larger)	12
PVC	4
Copper (2 in and larger)	10
Copper (1½ in and smaller	6

When installing cast-iron stacks, the base of each stack must be supported because of the weight of cast-iron pipe. See Figure 4.3 for an example of how to support the base of a cast-iron stack.

When installing pipe with flexible couplings, bands, or unions, the pipe must be installed and supported to prevent these flexible connections from moving. In larger pipes, pipes larger than 4 in, all flexible couplings must be supported to prevent the force of the pipe's flow from loosening the connection at changes in direction.

TABLE 4.17   Vertical Pipe-Support Intervals in Zone Two

Type of drainage pipe	Maximum distance of supports (ft)*
Lead pipe	4
Cast iron	At each story[†]
Galvanized	At each story[‡]
Copper (1¼ in and smaller)	4
Copper (1½ in and larger)	At each story
PVC (1½ in and smaller)	4
PVC (2 in and larger)	At each story
ABS (1½ in and smaller)	4
ABS (2 in and larger)	At each story

* All stacks must be supported at their bases.
† Support intervals may not exceed 15 ft.
‡ Support intervals may not exceed 30 ft.

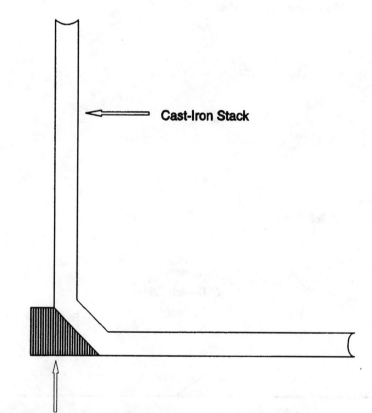

Masonry Block Or Equivalent

Figure 4.3 Support of a cast-iron stack.

Pipe-size reduction

As mentioned earlier, you may not reduce the size of a drainage pipe as it heads for the waste-disposal site. The pipe size may be enlarged, but it may not be reduced. There is one exception to this rule. Reducing closet bends, such as a 4-by-3 closet bend, are allowed.

Other facts to remember

A drainage pipe installed underground must have a minimum diameter of 2 in. When you are installing a horizontal branch fitting near the base of a stack, keep the branch fitting away from the point where the vertical stack turns to a horizontal run. The branch fitting should be installed at least 30 in back on a 3-in pipe and 40 in back on a 4-in pipe. By multiplying the size of the pipe by a factor of 10, you can determine how far back the branch fitting should be installed. See Fig. 4.4 for an example of this type of installation.

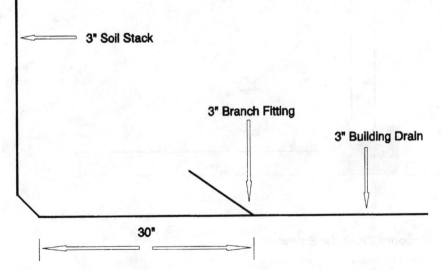

Figure 4.4 Example of the branch-fitting rule.

All drainage piping must be protected from the effects of flooding. When leaving a stub of pipe to connect with fixtures planned for the future, the stub must not be more than 2 ft in length and it must be capped. Some exceptions are possible on the prescribed length of a pipe stub. If you have a need for a longer stub, consult your local code officer. Cleanout extensions are not affected by the 2-ft rule.

Fittings

Fittings are also a part of the drainage system. Knowing when, where, and how to use the proper fittings is mandatory to the installation of a drainage system. Fittings are used to make branches and to change direction. The use of fittings to change direction is where we will start. When you wish to change direction with a pipe, you may have it change from a horizontal run to a vertical rise. You may be going from a vertical position to a horizontal one, or you might only want to offset the pipe in a horizontal run. Each of these three categories requires the use of different fittings. Let's take each circumstance and examine the fittings allowed.

Offsets in horizontal piping

When you want to change the direction of a horizontal pipe, you must use fittings approved for that purpose. You have six choices to choose from in zone three. Those choices are:

- Sixteenth bend (Fig. 4.5)
- Eighth bend (Fig. 4.6)
- Sixth bend (Fig. 4.7)
- Long-sweep fittings (Fig. 4.8)
- Combination wye and eighth bend (Fig. 4.9)
- Wye (Fig. 4.10)

Any of these fittings are generally approved for changing direction with horizontal piping, but as always, it is best to check with your local code officer for current regulations. For zone one's requirements of fittings used to change direction with horizontal piping, refer to Table 4.18.

Going from horizontal to vertical

You have a wider range of choice in selecting a fitting for going from a horizontal to a vertical position. There are nine possible candidates

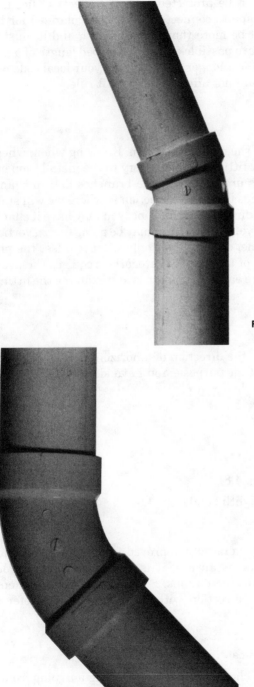

Figure 4.5 Sixteenth bend.

Figure 4.6 Eighth bend.

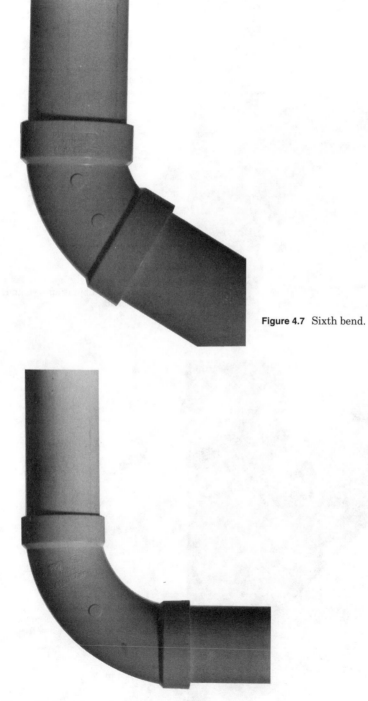

Figure 4.7 Sixth bend.

Figure 4.8 Long-sweep quarter bend.

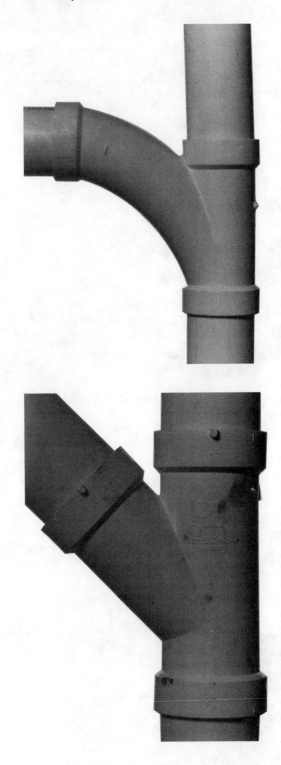

Figure 4.9 Combination wye and eighth bend.

Figure 4.10 Wye.

TABLE 4.18 Fittings Approved for Horizontal Changes in Zone One*

45° wye
Combination wye and eighth bend

* Other fittings with similar sweeps may also be approved.

available for this type of change in direction, when working in zone three. The choices are:

- Sixteenth bend
- Eighth bend
- Sixth bend
- Long-sweep fittings
- Combination wye and eighth bend
- Quarter bend (Fig. 4.11)
- Wye
- Short-sweep fittings
- Sanitary tee (Fig. 4.12)

You may not use a double sanitary tee in a back-to-back situation if the fixtures being served are of a blow-out or pump type. For example, you could not use a double sanitary tee to receive the discharge of two washing machines if the machines were positioned back to back. The sanitary tee's throat is not deep enough to keep drainage from feeding back and forth between the fittings. In a case like this, use a double combination wye and eighth bend. The combination fitting (Fig. 4.13) has a much longer throat and will prohibit waste water from transferring across the fitting to the other fixture.

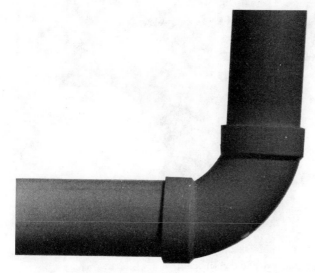

Figure 4.11 Quarter bend.

Figure 4.12 Sanitary tee.

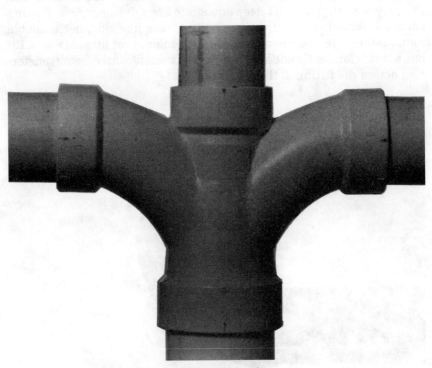

Figure 4.13 Double combination wye and eighth bend.

Vertical-to-horizontal changes in direction

There are seven fittings allowed to change direction from vertical to horizontal. These fittings are:

- Sixteenth bend
- Sixth bend
- Eighth bend
- Long-sweep fittings
- Wye
- Combination wye and eighth bend
- Short-sweep fittings that are 3-in or larger

For fittings allowed to change direction from horizontal to vertical in zone one, refer to Table 4.19. Table 4.20 shows the fittings allowed in zone one for changing direction from vertical to horizontal.

Zone one prohibits a fixture outlet connection within 8 ft of a vertical-to-horizontal change in direction of a stack, if the stack serves a suds-producing fixture. A suds-producing fixture could be a laundry fixture, a dishwasher, a bathing unit, or a kitchen sink. This rule does not apply to single-family homes and stacks in buildings with less than three stories.

Indirect Wastes

Indirect-waste requirements can pertain to a number of types of plumbing fixtures and equipment. These might include a clothes

---

TABLE 4.19   Fittings Approved for Horizontal-to-Vertical Changes in Zone One*

45° wye
60° wye
Combination wye and eighth bend
Sanitary tee
Sanitary tapped tee branches

---

* Cross-fittings, like double sanitary tees, cannot be used when they are of a short-sweep pattern; however, double sanitary tees can be used if the barrel of the tee is at least two pipe sizes larger than the largest inlet.

TABLE 4.20   Fittings Approved for Vertical-to-Horizontal Changes in Zone One

45° branches
60° branches and offsets if they are installed in a true vertical position

---

washer drain a condensate line, a sink drain, or the blow-off pipe from a relief valve, just to name a few. These indirect wastes are piped in this manner to prevent the possibility of contaminated matter backing up the drain into a potable water or food source, among other things.

Most indirect-waste receptors are trapped. If the drain from the fixture is more than 2 ft long, the indirect-waste receptor must be trapped. However, this trap rule applies to fixtures like sinks, not to an item such as a blow-off pipe from a relief valve. The rule is different in zone one. In zone one, if the drain is more than 4 ft long, it must be trapped.

The safest method for indirect waste is accomplished by using an air gap. When an air gap is used, the drain from the fixture terminates above the indirect-waste receptor, with open-air space between the waste receptor and the drain. This prevents any back-up or back-siphonage.

Some fixtures, depending on local code requirements, may be piped with an air break, rather than an air gap. With an air break, the drain may extend below the flood-level rim and terminate just above the trap's seal. The risk to an air break is the possibility of a back-up. Since the drain is run below the flood-level rim of the waste receptor, it is possible that the waste receptor could overflow and back up into the drain. This could create contamination, but in cases where contamination is likely, an air gap will be required. Check with your local codes office before using an air break.

Standpipes, like those used for washing machines, are a form of indirect-waste receptors. A standpipe used for this purpose in zones one and three must extend at least 18 in above the trap's seal, but they may not extend more than 30 in above the trap seal. If a clear-water waste receptor is located in a floor, zone three requires the lip of the receptor to extend at least 2 in above the floor. This eliminates the waste receptor from being used as a floor drain.

Choosing the proper size for a waste receptor is generally based on the receptor's ability to handle the discharge from a drain without excessive splashing. If you are concerned with sizing a particular waste receptor, talk with your local code officer for a ruling.

Buildings used for food preparation, storage, and similar activities are required to have their fixtures and equipment discharge drainage through an air gap. Zone three provides an exception to this rule. In zone three, dishwashers and open culinary sinks are excepted. Zone two requires the discharge pipe to terminate at least 2 in above the receptor. Zone one requires the distance to be a minimum of 1 in. Zones two and three require the air-gap distance to be a minimum of twice the size of the pipe discharging the waste. For example, a ½-in discharge pipe would require a 1-in air gap.

Zones two and three prohibit the installation of an indirect-waste receptor in any room containing toilet facilities. Zone one goes along with this ruling but allows one exception: the installation of a receptor for a clothes washer when the clothes washer is installed in the same room. Indirect-waste receptors are not allowed to be installed in closets and other unvented areas. Indirect-waste receptors must be accessible. Zone two requires all receptors to be equipped with a means of preventing solids with diameters of ½ in or larger from entering the drainage system. These straining devices must be removable to allow for cleaning.

When you are dealing with extreme water temperatures in waste water, such as with a commercial dishwasher, the dishwasher drain must be piped to an indirect waste. The indirect waste will be connected to the sanitary plumbing system, but the dishwasher drain may not connect to the sanitary system directly if the waste water temperature exceeds 140°F. Steam pipes may not be connected directly to a sanitary drainage system. Local regulations may require the use of special piping, sumps, or condensers to accept high-temperature water. Zone one prohibits the direct connection of any dishwasher to the sanitary drainage system.

Clear-water waste, from a potable source, must be piped to an indirect waste, with the use of an air gap. Sterilizers and swimming pools might provide two examples of when this rule would be used. Clear water from nonpotable sources, such as a drip from a piece of equipment, must be piped to an indirect waste. In zone three, an air break is allowed in place of an air gap. Zone two requires any waste entering the sanitary drainage system from an air conditioner to do so through an indirect waste.

Special Wastes

Special wastes are those wastes that may have a harming effect on a plumbing or waste-disposal system. Possible locations for special waste piping might include photographic labs, hospitals, or buildings where chemicals or other potentially dangerous wastes are dispersed. Small, personal-type photo darkrooms do not generally fall under the scrutiny of these regulations. Buildings that are considered to have a need for special waste plumbing are often required to have two plumbing systems, one system for normal sanitary discharge and a separate system for the special wastes. Before many special wastes are allowed to enter a sanitary drainage system, the wastes must be neutralized, diluted, or otherwise treated.

Depending upon the nature of the special wastes, special materials may be required. When you venture into the plumbing of special wastes, it is always best to consult the local code officer before proceeding with your work.

Sewer Pumps and Pits

There will be times when conditions will not allow a plumbing system to flow in the desired direction by gravity. When this is the case, sewer pumps and pits become involved. It is also possible that sump pumps and sumps will be used to remove water collected below the level of the building drain or sewer. When you plan to install a pump or sump pit, you must abide by certain regulations.

All sump pits must have a sealed cover that will not allow the escape of sewer gas. The pit size will be determined by the size and performance of the pump being housed in the sump. But, the sump generally must have a minimum diameter of 18 in and a minimum depth of 24 in. If a sewer pump is installed in a pit, the pump must be capable of lifting solids, with a diameter of 2 in, up into a gravity drain or sewer. The discharge pipe from these sumps must have a minimum diameter of 2 in. Zone one requires the sizing of the drain receiving the discharge from a sewer sump to be sized with a rating of 2 fixtures units for every gallon per minute the pump is capable of producing.

If the sump pit will not receive any discharge from toilets, the pump is not required to lift the 2-in solids and may be smaller. A standard procedure is to install a pump capable of lifting ½-in solids to a gravity drain if no toilets discharge into the sump.

It is a good idea to install two pumps in the sump. The pumps may be installed in a manner to take turns with the pumping chores, but most importantly, if one pump fails, the other pump can continue to operate. Zone two requires the installation of this type of two-pump system when six or more water closets discharge into the sump. Zone one requires any public installation of a sewer sump to be equipped with a two-pump system. Alarm systems are often installed on sewer pump systems. These alarms warn building occupants if the water level in the sewer pit rises to an unusually high level. Zones one and two require the effluent level to remain at least 2 in below the inlet of the sump.

All sewer sumps should be equipped with a vent. Ideally, the vent should extend upward to open air space without tying into another vent. Most sump vents are 2 in in diameter, but in no case shall they have a diameter of less than 1½ in.

There should be a check valve and a gate valve installed on the discharge piping from the pump. These devices prevent water from running back into the sump and allow the pump to be worked on with relative ease.

Most sewer pumps are equipped with a 2-in discharge outlet. An ejector pump with a 2-in outlet should be able to pump 21 gpm. If the discharge outlet is 3 in in diameter, the pump should have a flow rate of 46 gpm.

Storm-Water Drainage Piping

Storm-water drainage piping is a piping system designed to control and convey excess groundwater to a suitable location, which might be a catch basin, storm-sewer, or a pond. But, storm-water drainage may never be piped into a sanitary sewer or plumbing system. Up until now, our math requirements have been fairly simple, but that is about to change. Not that this section is a brain-buster, but it will require a little extra effort. The following sizing examples are based on zone three's requirements.

When you wish to size a storm-water drainage system, you must have some benchmark information to work with. One consideration is the amount of pitch a horizontal pipe will have on it. Another piece of the puzzle is the number of square feet of surface area your system will be required to drain. You will also need data on the rainfall rates in your area.

When you use your code book to size a storm-water system, you should have access to all the key elements required to size the job, except for the possibility of the local rainfall amounts. You should be able to obtain rainfall figures from your state or county offices. Your code book should provide you with a table to use in making your sizing calculations. Now, let's get into the computations needed to size a horizontal storm drain or sewer.

Sizing a horizontal storm drain or sewer

The first step to take when sizing a storm drain or sewer is to establish your known criteria. How much pitch will your pipe have on it? In the following example, the pipe will have a ¼-in/ft pitch. Knowing the pitch gives us a starting point and begins to take the edge off of an intimidating task.

Table 4.21 shows an example of a sizing table. To keep this process as simple as possible, it only includes a column in the table for our known grade of ¼ in/ft. In your code book you should be offered a few other choices, but since you are only going to use one pitch, once you know what it is, that is the only column that you need to pay attention to, as far as pitch is concerned.

TABLE 4.21   Example of a Horizontal Storm-Water Sizing Table*

Pipe grade (in/ft)	Pipe size (in)	Gallons per minute	Number of square feet of surface area
¼	3	48	4,640
¼	4	110	10,600
¼	6	314	18,880
¼	8	677	65,200

* These figures are based on a rainfall with a maximum rate of 1 in of rain per hour, for a full hour, and occurring once every 100 years.

What else do we know? Well, we know that the subject system is going to be located in Portland, Maine. Portland's rainfall is rated at 2.4 in/h. This rating assumes a 1-h storm that is only likely to occur once every 100 years. Now we have two of the factors needed to size our system.

We also know that the surface area that the system will be required to drain is 15,000 ft^2; this includes the roof and parking area. We're getting close to home now. We've got three of the elements needed to get this job done. But, how do we use the numbers in Table 4.21 to make any sense of this? Well, there are a couple of ways to ease the burden. When you are working with a standard table, like the ones found in most code books, you must convert the information to suit your local conditions. For example, if a standardized table is based on 1 in of rainfall an hour and your location has 2.4 in of rainfall per hour, you must convert the table, but this is not difficult.

When we want to convert a table based on a 1-in rainfall to meet local needs, all we have to do is divide the drainage area in the table by our rainfall amount. For example, if the standard chart shows an area of 10,000 ft^2 requiring a 4-in pipe, we can change the table by dividing our rainfall amount, 2.4, into the surface area of 10,000 ft^2.

If we divide 10,000 by 2.4, we get 4167. All of a sudden, we have solved the mystery of computing storm-water piping needs. With this simple conversion, we know that if our surface area was 4167 ft^2, we would need a 4-in pipe. But, our surface area is 15,000 ft^2, so, what size pipe do we need? Well, we know it will have to be larger than 4 in. So, we look down the conversion chart and find the appropriate surface area. Our 15,000 ft^2 of surface area will require a storm-water drain with a diameter of 8 in. We found this by dividing the surface areas of the numbers in Table 4.21 by 2.4 until we reached a number equal to, or greater than, our surface area. We could almost get by with a 6-in pipe, but not quite.

Now, let's recap this exercise. To size a horizontal storm drain or sewer, decide what pitch you will put on the pipe. Next, determine what your area's rainfall is for a 1-h storm, occurring each 100 years. If you live in a city, your city may be listed, with its rainfall amount, in your code book. Using a standardized chart, rated for 1 in of rainfall per hour, divide the surface area by a factor equal to your rainfall index; in the example it was 2.4. This division process converts a generic table into a customized table, just for your area.

Once the math is done, look down the table for the surface area that most closely matches the area you have to drain. To be safe, go with a number slightly higher than your projected number. It is better to have a pipe sized one size too large than one size too small. When you have found the appropriate surface area, look across the table to see what size pipe you need. See how easy that was. Well, maybe it's not easy, but it is a chore you can handle.

Sizing rain leaders and gutters

When you are required to size rain leaders or downspouts, you use the same procedure described above, with one exception. You use a table, supplied in your code book, to size the vertical piping. Determine the amount of surface area your leader will drain and use the appropriate table to establish your pipe size. The conversion factors are the same.

Sizing gutters is essentially the same as sizing horizontal storm drains. You will use a different table, provided in your code book, but the mechanics are the same.

Roof drains

Roof drains are often the starting point of a storm-water drainage system. As the name implies, roof drains are located on roofs. On most roofs, the roof drains are equipped with strainers that protrude upward, at least 4 in, to catch leaves and other debris. Roof drains should be at least twice the size of the piping connected to them. All roofs that do not drain to hanging gutters are required to have roof drains. A minimum of two roof drains should be installed on roofs with a surface area of 10,000 ft^2, or less. If the surface area exceeds 10,000 ft^2, a minimum of four roof drains should be installed.

When a roof is used for purposes other than just shelter, the roof drains may have a strainer that is flush with the roof's surface. Roof drains should obviously be sealed to prevent water from leaking around them. The size of the roof drain can be instrumental in the flow rates designed into a storm-water system. When a controlled flow from roof drains is wanted, the roof structure must be designed to accommodate the controlled flow.

More sizing information

If a combined storm-drain and sewer arrangement is approved, it must be sized properly. This requires converting fixture-unit loads into drainage surface area. For example, 256 fixture units will be treated as 1000 ft^2 of surface area. Each additional fixture unit, in excess of 256, will be assigned a value of 3%$_0$ ft^2. In the case of sizing for continuous flow, each gallon per minute is rated as 96 ft^2 of drainage area.

Some facts about storm-water piping

Storm-water piping requires the same amount of cleanouts, with the same frequency, as a sanitary system. Just as regular plumbing pipes must be protected, so shall storm-water piping. For example, if a downspout is in danger of being crushed by automobiles, you must install a guard to protect the downspout.

As stated earlier, storm-water systems and sanitary systems should not be combined. There may be some cities where the two are combined, but they are the exception rather than the rule. Area-way, or floor, drains must be trapped. When rain leaders and storm drains are allowed to connect to a sanitary sewer, they are required to be trapped. The trap must be equal in size to the drain it serves. Traps must be accessible for cleaning the drainage piping. Storm-water piping may not be used for conveying sanitary drainage.

Sump pumps

Sump pumps are used to remove water collected in building subdrains. These pumps must be placed in a sump, but the sump need not be covered with a gastight lid or vented. Many people are not sure what to do with the water pumped out of their basement by a sump pump. Do you pump it into your sewer? No, the discharge from a sump pump should not be pumped into a sanitary sewer. The water from the pump should be pumped to a storm-water drain or, in some cases, to a point on the property where it will not cause a problem.

All sump-pump discharge pipes should be equipped with a check valve. The check valve prevents previously pumped water from running down the discharge pipe and refilling the sump, forcing the pump to pull double duty. Here sump pumps are pumps removing groundwater, not waste or sewage.

Zone one's differences

Zone one has some additional requirements. See Tables 4.22 and 4.23 for zone one's requirements for storm-water materials. Note that once storm-water piping extends at least 2 ft from a building, any approved material may be used.

TABLE 4.22   Approved Materials for Storm-Water Drainage in Zone One for Use Inside Buildings, Above Ground

Galvanized
Wrought iron
Brass
Copper
Cast iron
ABS*
PVC*
Lead

* ABS and PVC may not be used in buildings that have more than three floors above grade.

TABLE 4.23   Approve Materials for Storm-Water Drainage in Zone One for Use Inside Buildings, Below Ground

Service-weight cast iron
DWV copper
ABS
PVC
Extra-strength vitrified clay

With zone one, the inlet area of a roof drain is generally only required to be 1½ times the size of the piping connected to the roof drain. However, when positioned on roofs used for purposes other than weather protection, roof drain openings must be sized to be twice as large as the drain connecting to them.

Zone one also provides tables for sizing purposes. When computing the drainage area, you must take into account the effect vertical walls will have on the drainage area. For example, a vertical wall that reflects water onto the drainage area must be allowed for in your surface-area computations. In the case of a single vertical wall, add one-half of the wall's total square footage to the surface area.

Two vertical walls that are adjacent to each other require you to add 35 percent of the combined wall square footage to your surface area.

If you have two walls of the same height that are opposite of each other, no added space is needed. In this case, each wall protects the other one and does not allow extra water to collect on the roof area.

When you have two opposing walls with different heights, you must make a surface-area adjustment. Take the square footage of the highest wall, as it extends above the other wall, and add half of the square footage to your surface area.

When you encounter three walls, you use a combination of the above instructions to reach your goal. Four walls of equal height do not require an adjustment. If the walls are not of equal height, use the procedures listed above to compute your surface area.

Zone two's differences

It would be nice if all plumbing codes were the same, but they are not. The following information provides insight into how zone two varies from zones one and three. See Tables 4.24, 4.25, and 4.26 for material requirements in zone two.

Sump pits are required to have a minimum diameter of 18 in. Floor drains may not connect to drains intended solely for storm water. When computing surface area to be drained for vertical walls, such as walls enclosing a roof-top stairway, use one-half of the total square

TABLE 4.24   Approved Materials for Storm-Water Drainage in Zone Two for Underground Use

Cast iron
Coated aluminum
ABS*
PVC*
Copper*
Concrete*
Asbestos-Cement*
Vitrified clay*

* These materials may be allowed for use, subject to local code authorities.

TABLE 4.25   Approved Materials for Storm-Water Drainage in Zone Two for Building Storm Sewers

Cast iron
Aluminum*
ABS
PVC
Vitrified clay
Concrete
Asbestos-cement

* Buried aluminum must be coated.

TABLE 4.26   Approved Materials for Storm-Water Drainage in Zone Two for Above-Ground Use

Galvanized
Black steel
Brass
DWV thicker types of copper
Cast iron
ABS
PVC
Aluminum
Lead

footage from the vertical wall surface that reflects water onto the drainage surface.

Some roof designs require a back-up drainage system for emergencies. These roofs are generally roofs that are surrounded by vertical sections. If these vertical sections are capable of retaining water on the roof if the primary drainage system fails, a secondary drainage system is required. In these cases, the secondary system must have independent piping and discharge locations. These special systems are sized with the use of different rainfall rates. The ratings are based on a 15-minute (min) rainfall. Otherwise, the 100-year conditions still apply.

Zone two's requirements for sizing a continuous flow require a rating of 24 ft^2 of surface area to be given for every gallon per minute generated. For regular sizing, based on 4 in of rain per hour, 256 fixture units equal 1000 ft^2 of surface area. Each additional fixture unit is rated at 3%₁₀ in. If the rainfall rate varies, a conversion must be done.

To convert the fixture-unit ratings to a higher or lower rainfall, you must do some math. Take the square foot rating assigned to fixture units and multiply it by 4. For example, 256 fixture units equal 1000 ft^2. Multiply 1000 by 4, and get 4000. Now, divide the 4000 by the rate of rainfall for 1 h. Say for example that the hourly rainfall was 2 in; the converted surface area would be 2000.

Well, you have made it past a section of code regulations that gives professional plumbers the most trouble. Storm-water drains are despised by some plumbers because they have little knowledge of how to compute them. With the aid of this chapter, you should be able to design a suitable system with minimal effort.

Reminder Notes

Zone one

1. Fixture-unit loads on traps are different from those found in zones two and three. Refer to Table 4.3 for zone one's requirements.

2. A 2-in horizontal drain may carry up to eight fixture units, with exceptions.

3. A 3-in horizontal drain may carry up to 35 fixture units and of these units as many as three water closets may be installed.

4. A 1½-in horizontal drain may carry up to two fixture units, except it may not carry the drainage of sinks, urinals, and dishwashers.

5. A 4-in horizontal drain is rated for up to 216 fixture units.

6. Special permission is required to install pipe with a ⅛-in/ft grade.

7. See Table 4.14 for pipe support intervals of horizontal piping.

8. See Table 4.15 for pipe support intervals of vertical piping.

9. See Table 4.18 for fittings allowed to change the direction of horizontal piping.

10. See Table 4.20 for fittings allowed to change from a vertical to a horizontal direction.

11. See Table 4.19 for fittings allowed to change from a horizontal to a vertical direction.

12. In buildings, other than single-family homes and buildings with less than three stories, a fixture outlet may not connect to a stack within 8 ft of a vertical to horizontal change of direction, when the stack is receiving the discharge of a suds-producing fixture.

13. An indirect-waste drain that is 5 ft long, or longer, must be trapped.

14. Indirect-waste receptors may not be installed in rooms with toilet facilities, except for washing-machine receptors, when the clothes washer is installed in the same room.

15. No dishwasher is allowed to be connected directly to a sanitary drainage system.

16. Indirect-waste pipes in buildings used for food preparation, and similar activities, must terminate at least 2 in above the indirect-waste receptor.

17. Refrigerators that are used for the storage of prepackaged goods, like bottles of soda, are excepted from the indirect-waste rules.

18. Air conditioning equipment may be piped with an air break, but food-related equipment and fixtures must be piped with air gaps.

19. A vent from an indirect waste may not tie into a vent that is connected to a sewer.

20. Fixtures that produce waste under pressure are required to be piped to indirect-waste receptors.

21. Condensate drains from air conditioning coils may be connected directly to a lavatory tailpiece or tub waste, under special conditions. The connection point must be accessible and in a place controlled by the same person controlling the air conditioner.

22. Pure condensate, from a fuel-burning appliance, being discharged into a drainage system must only discharge into drainage systems constructed of materials approved for this purpose.

23. It is permissible to discharge the drainage from a watercooler to an indirect waste.

24. Owners of buildings containing chemical-waste piping must make and keep a detailed record of the location of all piping in these systems.

25. Chemical-waste pipes should be installed in a way to make them as readily accessible as reasonably possible.

26. Drains accepting the waste for sewer sumps must be sized with a rating of 2 fixture units for every gallon per minute the pump is capable of producing.

27. Any installation of a sewer sump that will serve the public requires the installation of a two-pump system.

28. Effluent levels in sewer pump sumps must not rise to a level closer than 2 in to the sump inlet.

29. Sewer sump vents when serving an air-operated sewer ejector pump may not connect to other vents.

30. Water-operated sewer ejectors are not approved for use.

Zone two

1. See Table 4.16 for pipe support intervals of horizontal piping.

2. See Table 4.17 for pipe support intervals of vertical piping.

3. Indirect-waste pipes in buildings used for food preparation and similar activities, must terminate at least 2 in above the indirect-waste receptor.

4. Any waste from an air conditioner that discharges into the sanitary drainage system must do so through an indirect waste.

5. Indirect-waste receptors must be equipped with a means of preventing solids with diameters of ½ in or larger from entering the drainage system. The straining devices must be able to be removed for cleaning.

6. It is permissible to discharge the drainage from a watercooler to an indirect waste.

7. Sewer ejector sumps are required to be equipped with a duplex pumping system when six or more water closets discharge into the sump.

8. Effluent levels in sewer pump sumps must not rise to a level closer than 2 in to the sump inlet.

Zone three

1. Fixture-unit loads for a continuous flow are assigned at a rate of 1 fixture unit for every 7½ gpm.

2. Pipes with diameters in excess of 8 in can be installed with a grade of ¹⁄₁₆ in/ft.

3. Indirect-waste receptors for clear-water wastes that are in a floor must extend at least 2 in above the floor level.

4. Zone three does not require dishwashers and open culinary sinks to be piped to an indirect waste.

5. Clear-water wastes from nonpotable sources may be piped to an indirect waste by using an air break.

Understanding the Vent System

Most people don't think much about vents when they consider the plumbing in their homes or offices, but vents play a vital role in the scheme of sanitary plumbing. Many plumbers underestimate the importance of vents. The sizing and installation of vents often cause more confusion than the same tasks applied to drains. This chapter will teach you the role and importance of vents. It will also instruct you in the proper methods of sizing and installing them.

Whether you are working with simple individual vents or complex island vents, this chapter will improve your understanding and installation of them. Why do we need vents? They perform three easily identified functions. The most obvious function of a vent is its capacity to carry sewer gas out of a building and into the open air. A less obvious, but equally important, aspect of the vent is its ability to protect the seal in the trap it serves. The third characteristic of the vent is its ability to enable drains to drain faster and better. Let's look more closely at each of these factors.

Transportation of Sewer Gas

Vents transport sewer gas through a building, without exposing occupants of the building to the gas, to an open air space. Why is this important? Sewer gas can cause health problems. The effect of sewer gas on individuals will vary, but it should be avoided by all individuals. In addition to health problems caused by sewer gas, explosions are also possible when sewer gas is concentrated in a poorly ventilated area. Yes, sewer gas can create an explosion when it is concentrated, confined, and ignited. As you can see, just from looking at this single purpose of vents, vents are an important element of a plumbing system.

Protecting Trap Seals

Another job plumbing vents perform is the protection of trap seals. The water sitting in a fixture's trap blocks the path of sewer gas trying to enter the plumbing fixture. Without a trap seal, sewer gas could rise through the drainage pipe and enter a building through a plumbing fixture. As mentioned above, this could result in health problems and the risk of explosion. Good trap seals are essential to sanitary plumbing systems.

Vents protect trap seals by regulating the atmospheric pressure applied to them. It is possible for pressures to rise in unvented traps to a point where the trap contents actually expel into the fixture it serves. This is not a common problem, but if it occurs, the plumbing fixture could become contaminated.

A more likely problem is when the pressure on a trap seal is reduced and becomes something of a vacuum. When this happens, the water creating the trap seal is sucked out of the trap and down the drain. Once the water is taken from the trap, there is no trap seal. The trap will remain unsealed until water is replaced in the trap. Without water in it, a trap is all but useless. Vents prevent these extreme atmospheric pressure changes, therefore, protecting the trap seal.

Tiny Tornados

Have you ever drained your sink or bathtub and watched the tiny water tornados? When you see the fast swirling action of water being pulled down a drain, it usually indicates that the drain is well vented. If water is sluggish and moves out of the fixture like a lazy river, the vent for the fixture, if there is one, is not performing at its best.

Vents help fixtures to drain faster. The air allowed from the vent keeps the water moving at a more rapid pace. This not only entertains us with tiny tornados, but it aids in the prevention of clogged pipes. It is possible for drains to drain too quickly, removing the liquids and leaving hair, grease, and other potential pipe blockers present. However, if a pipe is properly graded and does not contain extreme vertical drops into improper fittings, such problems should not occur.

Do All Plumbing Fixtures Have Vents?

Most local plumbing codes require all fixture traps to be vented, but there are exceptions. In some jurisdictions, combination waste and vent systems are used. In a combination waste and vent system, vertical vents are rare. Instead of vertical vents being used, larger drainage pipes are used. The larger diameter of the drain allows air to circulate in the pipe, eliminating the need for a vent, as far as satisfactory drainage is concerned. Experience with both types of systems has shown that vented systems perform much better than combination waste and vent systems.

Combination waste and vent systems do not have vents on each fixture, so how is the trap seal protected? Trap seals in a combination waste and vent system are protected through the use of antisyphon traps or drum traps. Vented systems normally use P-traps. By using an antisiphon or drum trap, the trap is not susceptible to back-siphonage. Since these traps are larger, deeper, and made so that the water in the trap is not replaced by fresh water with each use of the fixture, they are not required to be vented, subject to local code requirements.

Most jurisdictions prohibit the use of drum traps and require traps to be vented. Before you install your plumbing, check with the local code officer for the facts pertinent to your location. The following tables show piping requirements for each zone. Table 5.1 shows approved above-ground vent materials for zone one. Table 5.2 does the same for zone two, and Table 5.3 covers approved vent materials for zone three. Table 5.4 gives a listing of approved underground venting materials for zone one. Table 5.5 covers underground materials approved for use in zone two. Table 5.6 shows approved underground materials for zone three. Fittings for vent piping must be compatible with the piping used.

TABLE 5.1   Materials Approved for Above-Ground Vents in Zone One

Cast iron
ABS*
PVS*
Copper
Galvanized
Lead
Brass

* These materials may not be used with buildings having more than three floors above grade.

TABLE 5.2   Materials Approved for Above-Ground Vents in Zone Two

Cast iron
ABS
PVC
Copper
Galvanized
Lead
Aluminum
Borosilicate Glass
Brass

TABLE 5.3   Materials Approved for Above-Ground Vents in Zone Three

Cast iron
ABS
PVC
Copper
Galvanized
Lead
Aluminum
Brass

TABLE 5.4   Materials Approved for Underground Vents in Zone One

Cast iron
ABS*
PVC*
Copper
Brass
Lead

* These materials may not be used with buildings having more than three floors above grade.

TABLE 5.5   Materials Approved for Underground Vents in Zone Two

Cast iron
ABS
PVC
Copper
Aluminum
Borosilicate glass

TABLE 5.6   Materials Approved for Underground Vents in Zone Three

Cast iron
ABS
PVC
Copper

Individual Vents

Individual vents are, as the name implies, vents that serve individual fixtures. These vents only vent one fixture, but they may connect into another vent that will extend to the open air. Individual vents do not have to extend from the fixture being served to the outside air, without joining another part of the venting system, but they must vent to open air space. See Fig. 5.1 for an example of an individual vent.

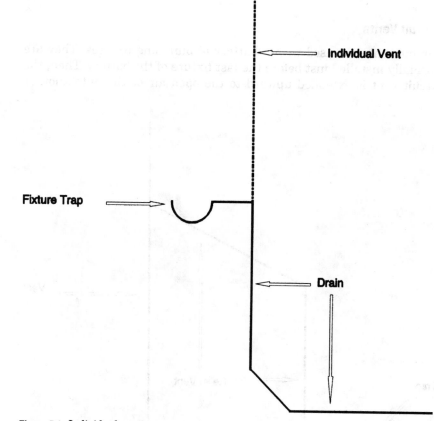

Figure 5.1 Individual vent.

Sizing an individual vent is easy. The vent must be at least one-half the size of the drain it serves, but it may not have a diameter of less than 1¼ in. For example, a vent for a 3-in drain could, in most cases, have a diameter of 1½ in. A vent for a 1½-in drain may not have a diameter of less than 1¼ in.

Relief Vents

Relief vents are used in conjunction with other vents. Their purpose is to provide additional air to the drainage system when the primary vent is too far from the fixture. See Fig. 5.2 for an example of a relief vent. Relief vents must be at least one-half the size of the pipe it is venting. For example, if a relief vent is venting a 3-in pipe, the relief vent must have a 1½-in or larger diameter.

Circuit Vents

Circuit vents are used with a battery of plumbing fixtures. They are normally installed just before the last fixture of the battery. Then, the circuit vent is extended upward to the open air or tied into another

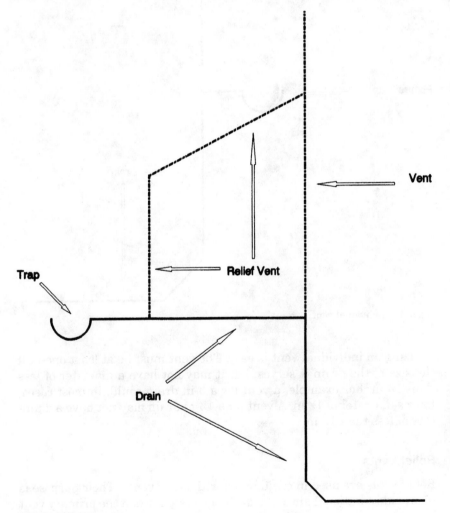

Figure 5.2 Relief vent.

vent that extends to the outside. Circuit vents may tie into stack vents or vent stacks. See Fig. 5.3 for an example of a circuit vent. When sizing a circuit vent, you must account for its developed length. But in any event, the diameter of a circuit vent must be at least one-half the size of the drain it is serving.

Vent Sizing Using Developed Length

What effect does the length of the vent have on the vent's size? The developed length, the total linear footage of pipe making up the vent, is

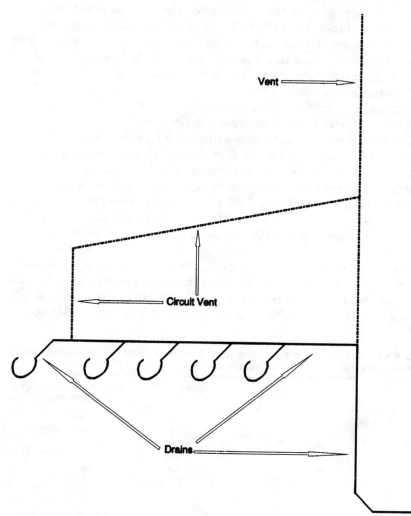

Figure 5.3 Circuit vent.

used in conjunction with factors provided in code books to determine vent sizes. To size circuit vents, branch vents, and individual vents for horizontal drains, you must use this method of sizing.

The criteria needed for sizing a vent, based on developed length, are the grade of the drainage pipe, the size of the drainage pipe, the developed length of the vent, and the factors allowed by local code requirements. Let's look at a few examples of how to size a vent using this method.

For our first example, assume the drain you are venting is a 3-in pipe with a ¼-in/ft grade. This sizing exercise is done using zone three requirements. Look at Table 5.7 and find the proper pipe size and grade. Now, looking at the table, notice the number listed under the 1½-in vent column. You will see the number is 97. This means that a 3-in drain, running horizontally, with a ¼-in/ft grade, can be vented with 1½-in vent that has a developed length of 97 ft. It would be rare to extend a vent anywhere near 97 ft, but if your vent needed to exceed this distance, you could go to a larger vent. A 2-in vent would allow you to extend the vent for a total length of 420 ft. A vent larger than 2 in would allow you to extend the vent indefinitely.

For the second example, still using zone three's rules, assume the drain is a 4-in pipe, with a ¼-in/ft grade. In this case you could not use a 1½-in vent. Remember, the vent must be at least one-half the size of the drain it is venting. A 2-in vent would allow a developed vent length of 98 ft, and a 3-in vent would allow the vent to extend to an unlimited length. As you can see, this type of sizing is not difficult.

Now, let's size a vent with the use of zone one's rules. In zone one, vent sizing is based on the vent's length and the number of fixture units on the vent. If you were sizing a vent for a lavatory, you would need to know how many fixture units the lavatory represents. Lavatories are rated as 1 fixture unit. By using a table in the code book, you would find that a vent serving 1 fixture unit can have a diameter of

TABLE 5.7 Vent Sizing Table for Zone Three (For Use with Individual, Branch, and Circuit Vents for Horizontal Drain Pipes)

Drain pipe size (in)	Drain pipe grade (in/ft)	Vent pipe size (in)	Maximum developed length of vent pipe (ft)
1½	¼	1¼	Unlimited
1½	¼	1½	Unlimited
2	¼	1¼	290
2	¼	1½	Unlimited
3	¼	1½	97
3	¼	2	420
3	¼	3	Unlimited
4	¼	2	98
4	¼	3	Unlimited
4	¼	4	Unlimited

1¼ in and extend for 45 ft. A bathtub, rated at 2 fixture units, would require a 1½-in vent. The bathtub vent could run for 60 ft.

Branch Vents

Branch vents are vents extending horizontally that connect multiple vents together. Figure 5.4 shows an example of a branch vent. Branch vents are sized with the developed-length method, just as you were shown in the examples above. A branch vent or individual vent that is the same size as the drain it serves is unlimited in the developed length it may obtain. Be advised, zone two and zone three use different

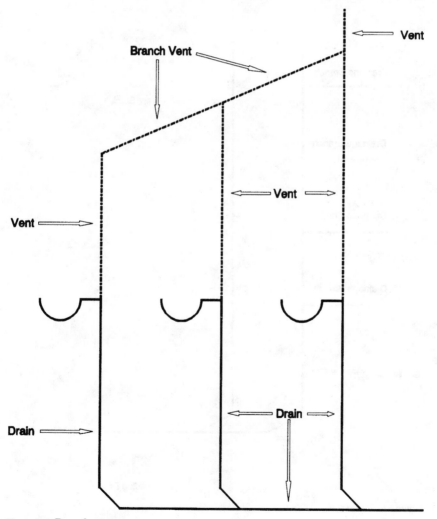

Figure 5.4 Branch vent.

tables and ratings for sizing various types of vents; zone one uses the same rating and table for all normal venting situations.

Vent Stacks

A vent stack is a pipe used only for the purpose of venting. Vent stacks extend upward from the drainage piping to the open air, outside of a building. Vent stacks are used as connection points for other vents, such as branch vents. A vent stack is a primary vent that accepts the connection of other vents and vents an entire system. Refer to Fig. 5.5

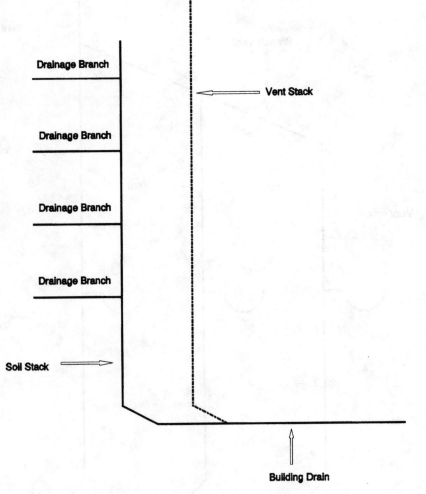

Figure 5.5 Vent stack.

for an example of a vent stack. Vent stacks run vertically and are sized a little differently.

The basic procedure for sizing a vent stack is similar to that used with branch vents, but there are some differences. Refer to Table 5.8 for an example of the criteria needed to size a vent stack in zone three. Zone two uses a very similar table, but the numbers vary in some instances. You must know the size of the soil stack, the number of fixture units carried by the soil stack, and the developed length of your vent stack. With this information and the regulations of your local plumbing code, you can size your vent stack. Let's work on an example.

Assume your system has a soil stack with a diameter of 4 in. This stack is loaded with 43 fixture units. Your vent stack will have a developed length of 50 ft. What size pipe will you have to use for your vent stack? When you look at the table, you will see that a 2-in pipe, used as a vent for the described soil stack, would allow a developed length of 35 ft. Your vent will have a developed length of 50 ft, so you can rule out 2-in pipe. In the column for 2½-in pipe, you see a rating for up to 85 ft. Since your vent is only going 50 ft, you could use a 2½-in pipe. However,

TABLE 5.8   Vent Sizing Table for Zone Three (For Use with Vent Stacks and Stack Vents)

Drain pipe size (in)	Fixture-unit load on drain pipe	Vent pipe size (in)	Maximum developed length of vent pipe (ft)
1½	8	1¼	50
1½	8	1½	150
1½	10	1¼	30
1½	10	1½	100
2	12	1½	75
2	12	2	200
2	20	1½	50
2	20	2	150
3	10	1½	42
3	10	2	150
3	10	3	1040
3	21	1½	32
3	21	2	110
3	21	3	810
3	102	1½	25
3	102	2	86
3	102	3	620
4	43	2	35
4	43	3	250
4	43	4	980
4	540	2	21
4	540	3	150
4	540	4	580

since 2½-in pipe is not common, you would probably use a 3-in pipe. This same sizing method is used when computing the size of stack vents.

Stack Vents

Stack vents are really two pipes in one. The lower portion of the pipe is a soil pipe, and the upper portion is a vent. This is the type of primary vent most often found in residential plumbing. Figure 5.6 shows you

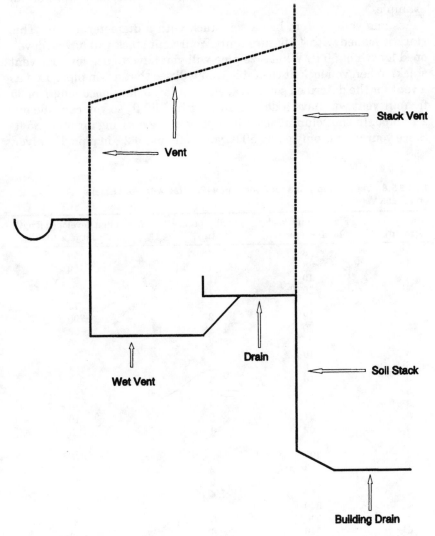

Figure 5.6 Stack vent.

what a stack vent looks like. Stack vents are sized with the same methods used on vent stacks.

Common Vents

Common vents are single vents that vent multiple traps. Figure 5.7 shows a diagram of a typical common vent. Common vents are only allowed when the fixtures being served by the single vent are on the

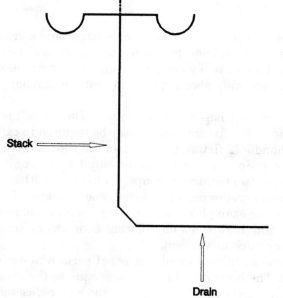

Figure 5.7 Common vent.

same floor level. Zone one requires the drainage of fixtures being vented with a common vent to enter the drainage system at the same level. Normally, not more than two traps can share a common vent, but there is an exception in zone three. Zone three allows you to vent the traps of up to three lavatories with a single common vent. Common vents are sized with the same technique applied to individual vents.

Island Vents

Island vents are unusual looking vents. They are allowed for use with sinks and lavatories. The primary use for these vents is with the trap of a kitchen sink, when the sink is placed in an island cabinet. Sometimes pictures speak louder than words; refer to Fig. 5.8 for a detail of how island venting works.

As you can see from the figure, island venting may take a little getting used to. Notice that the vent must rise as high as possible under the cabinet before it takes a U-turn and heads back downward. Since this piping does not rise above the flood-level rim of the fixture, it must be considered a drain. Fittings approved for drainage must be used in making an island vent.

The vent portion of an island vent must be equipped with a cleanout. The vent may not tie into a regular vent until it rises at least 6 in above the flood-level rim of the fixture.

Wet Vents

Wet vents are pipes that serve as a vent for one fixture and a drain for another. Wet vents, once you know how to use them, can save you a lot of money and time. By effectively using wet vents you can reduce the amount of pipe, fittings, and labor required to vent a bathroom group or two.

The sizing of wet vents is based on fixture units. The size of the pipe is determined by how may fixture units it may be required to carry. A 3-in wet vent can handle 12 fixture units. A 2-in wet vent is rated for 4 fixture units, and a 1½-in wet vent is allowed only 1 fixture unit. It is acceptable to wet vent two bathroom groups, six fixtures, with a single vent, but the bathroom groups must be on the same floor level. Figures 5.9 and 5.10 show some examples of wet venting. Zone two makes provisions for wet venting bathrooms on different floor levels. Zone one takes a different approach to wet venting.

Zone two has some additional regulations that pertain to wet venting; here they are. The horizontal branch connecting to the drainage stack must enter at a level equal to, or below, the water-closet drain. However, the branch may connect to the drainage at the closet bend.

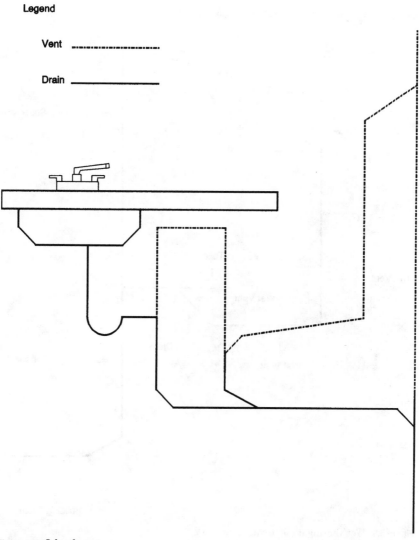

Legend

Vent   ----------------------------

Drain   _____

Figure 5.8 Island vent.

When wet venting two bathroom groups, the wet vent must have a minimum diameter of 2 in.

Table 5.9 shows the ratings used to size a wet-vented stack. Kitchen sinks and washing machines may not be drained into a 2-in combination waste and vent. Water closets and urinals are restricted on vertical combination waste and vent systems.

As for zone two's allowance in wet venting on different levels, here are the facts. Wet vents must have at least a 2-in diameter. Water clos-

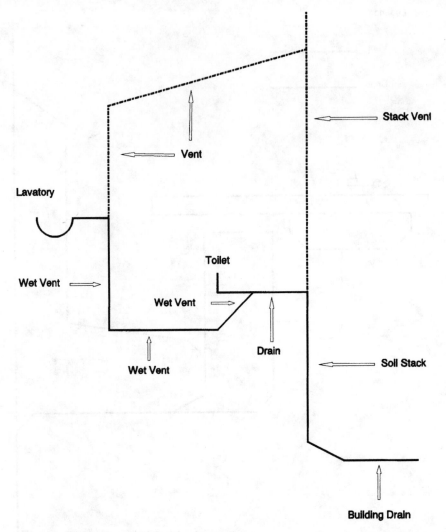

Figure 5.9 Wet venting a toilet with a lavatory.

ets that are not located on the highest floor must be back-vented. If, however, the wet vent is connected directly to the closet bend, with a 45° bend, the toilet being connected is not required to be back-vented, even if it is on a lower floor. Table 5.10 shows the ratings used to size a vent stack for a wet-vented application.

Zone one limits wet venting to vertical piping. These vertical pipes are restricted to receiving only the waste from fixtures with fixture-unit ratings of 2, or less, and that serve to vent no more than four fix-

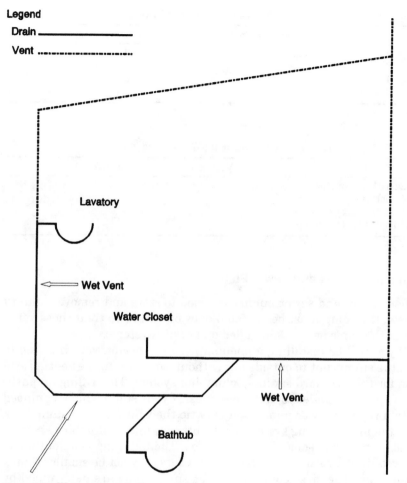

Figure 5.10 Wet venting a bathroom group.

tures. Wet vents must be one pipe size larger than normally required, but they must never be smaller than 2 in in diameter.

Crown Vents

A crown vent is a vent that extends upward from a trap or trap arm. Crown-vented traps are not allowed. When crown vents are used, they are normally used on trap arms, but even then, they are not common. Figure 5.11 shows how an approved crown vent would look. The vent must be on the trap arm, and it must be behind the trap by a distance equal to twice the pipe size. For example, on a 1½-in trap, the crown vent would have to be 3 in behind the trap, on the trap arm.

TABLE 5.9   Table for Sizing a Wet Stack Vent in Zone Two

Stack pipe size	Fixture-unit load on stack	Maximum length of stack (ft)
2	4	30
3	24	50
4	50	100
6	100	300

TABLE 5.10   Table for Sizing a Vent Stack for Wet Venting in Zone Two

No. of fixtures	Vent-stack size requirements (in)
1–2 bathtubs or showers	2
3–5 bathtubs or showers	2½
6–9 bathtubs or showers	3
10–16 bathtubs or showers	4

Vents for Sumps and Sewer Pumps

When sumps and sewer pumps are used to store and remove sanitary waste, the sump must be vented. Zones one and two treat these vents about the same as vents installed on gravity systems.

If you will be installing a pneumatic sewer ejector, you will need to run the sump vent to outside air, without tying it into the venting system for the standard sanitary plumbing system. This ruling on pneumatic pumps applies to all three zones. If your sump will be equipped with a regular sewer pump, you may tie the vent from the sump back into the main venting system for the other sanitary plumbing.

Zone three has some additional rules. The following is an outline of the requirements in zone three. Sump vents may not be smaller than a 1¼-in pipe. The size requirements for sump vents are determined by the discharge of the pump. For example, a sewer pump capable of producing 20 gpm could have its sump vented for an unlimited distance with a 1½-in pipe. If the pump was capable of producing 60 gpm, a 1½-in pipe could not have a developed length of more than 75 ft.

In most cases, a 2-in vent is used on sumps, and the distance allowed for developed length is not a problem. However, if your pump will pump more than 100 gpm, you had better take the time to do some math. Your code book will provide you with the factors you need to size your vent, and the sizing is easy. You simply look for the maximum discharge capacity of your pump and match it with a vent that allows the developed length you need.

This concludes the general description and sizing techniques for various vents. Next, we are going to look at regulations dealing with the methods of installation for vents.

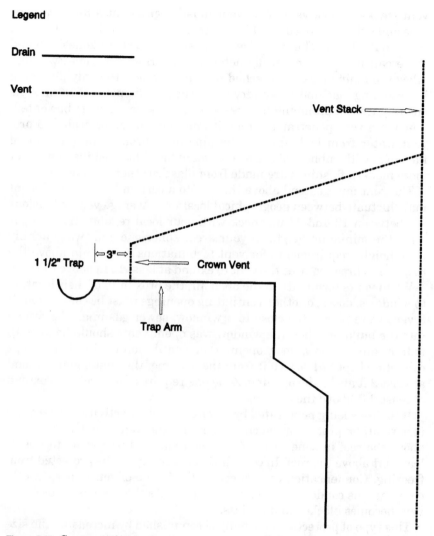

Figure 5.11 Crown venting.

Vent-Installation Requirements

Since there are so many types of vents and their role in the plumbing system is so important, there are many regulations affecting the installation of vents. What follows are specifics for installing various vents.

In zone two, any building equipped with plumbing must also be equipped with a main vent. Zone three requires any plumbing system that receives the discharge from a water closet to have either a main vent stack or stack vent. This vent must originate at a 3-in drainage pipe and extend upward until it penetrates the roof of the building and meets outside air. The vent size requirements for both zones two and three call for a minimum diameter of 3 in. However, zone two does allow the main stack in detached buildings, where the only plumbing is a washing machine or laundry tub, to have a diameter of 1½ in. Zone one requires all plumbing fixtures, except for exceptions, to be vented.

When a vent penetrates a roof, it must be flashed or sealed to prevent water from leaking past the pipe and through the roof. Metal flashings with rubber collars are normally used for flashing vents, but more modern flashings are made from plastic rather than metal.

The vent must extend above the roof to a certain height. The height may fluctuate between geographical locations. Average vent extensions are between 12 and 24 in; check with your local regulations to determine the minimum height in your area. Zones one and two generally have height requirements for vent terminations set at 6 in above the roof. Zone three requires the vent to extend at least 12 in above the roof.

When vents terminate in the open air, the proximity of their location to windows, doors, or other ventilating openings must be considered. If a vent were placed too close to a window, sewer gas might be drawn into the building when the window was open. Vents should be kept 10 ft from any window, door, opening, or ventilation device. If the vent cannot be kept at least 10 ft from the opening, the vent should extend at least 2 ft above the opening. Zone one requires these vents to extend at least 3 ft above the opening.

If the roof being penetrated by a vent is used for activities other than just weather protection, such as a patio, the vent must extend 7 ft above the roof in zone three. Zone two requires these vents to rise at least 5 ft above the roof. In cold climates, vents must be protected from freezing. Condensation can collect on the inside of vent pipes. In cold climates this condensation may turn to ice. As the ice mass grows, the vent becomes blocked and useless.

This type of protection is usually accomplished by increasing the size of the vent pipe. This ruling normally applies only in areas where temperatures are expected to be below 0°F. Zone three requires vents in this category to have a minimum diameter of 3 in. If this requires an increase in pipe size, the increase must be made at least 1 ft below the roof. In the case of side-wall vents, the change must be made at least 1 ft inside the wall.

Zone one's rules for protecting vents from frost and snow are a little different. All vents must have diameters of at least 2 in but never less than the normally required vent size. Any change in pipe size must take place at least 12 in before the vent penetrates into open air, and the vent must extend to a height of 10 in.

There may be occasions when it is better to terminate a plumbing vent out the side of a wall rather than through a roof. Zone one prohibits side-wall venting. Zone two prohibits side-wall vents from terminating under any building's overhang. When side-wall vents are installed, they must be protected against birds and rodents with a wire mesh or similar cover. Side-wall vents must not extend closer than 10 ft to the property boundary of the building lot. If the building is equipped with soffit vents, side-wall vents may not be used if they terminate under the soffit vents. This rule is in effect to prevent sewer gas from being sucked into the attic of the home.

Zone three requires buildings having soil stacks with more than five branch intervals to be equipped with a vent stack. Zone one requires a vent stack with buildings having at least 10 stories above the building drain. The vent stack will normally run up near the soil stack. The vent stack must connect into the building drain at or below the lowest branch interval. Figure 5.12 shows you an example of an approved vent stack installation. The vent stack must be sized according to the instructions given earlier. In zone three, the vent stack must be connected within 10 times its pipe size on the downward side of the soil stack. This means that a 3-in vent stack must be within 30 in of the soil stack on the downward side of the building drain.

Zone one further requires these stack vents to be connected to the drainage stack at intervals of every five stories. The connection must be made with a relief yoke vent. The yoke vent (Fig. 5.13) must be at least as large as either the vent stack or soil stack, whichever is smaller. This connection must be made with a wye fitting that is at least 42 in off the floor.

In large plumbing jobs, where there are numerous branch intervals, it may be necessary to vent offsets in the soil stack. Normally, the offset must be more than 45° to warrant an offset vent. Zones two and three require offset vents when the soil stack offsets and has five or more branch intervals above it. See Fig. 5.14 for an example of this procedure in zone three.

Just as drains are installed with a downward pitch, vents must also be installed with a consistent grade (Fig. 5.15). Vents should be graded to allow any water entering the vent pipe to drain into the drainage system. A typical grade for vent piping is ¼-in/ft. Zone one allows vent pipes to be installed level, without pitch (Fig. 5.16).

Dry vents must be installed in a manner to prevent clogging and blockages. You may not lay a fitting on its side and use a quarter bend to turn the vent up vertically. Dry vents should leave the drainage pipe in a vertical position. An easy way to remember this is that if you need an elbow to get the vent up from the drainage, you are doing it the wrong way.

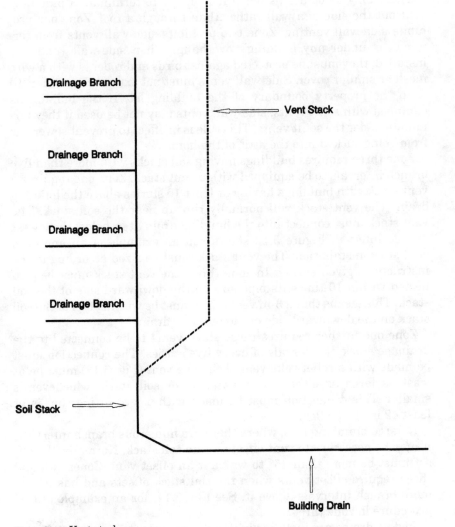

Figure 5.12 Vent stack.

Most vents can be tied into other vents, such as a vent stack or stack vent. But the connection for the tie-in must be at least 6 in above the flood-level rim of the highest fixture served by the vent.

Zone two allows the use of circuit vents to vent fixtures in a battery. The drain serving the battery must be operating at one-half of its fixture-unit rating. If the application is on a lower-floor battery with a minimum of three fixtures, relief vents are required. You must also pay attention to the fixtures draining above these lower-floor batteries.

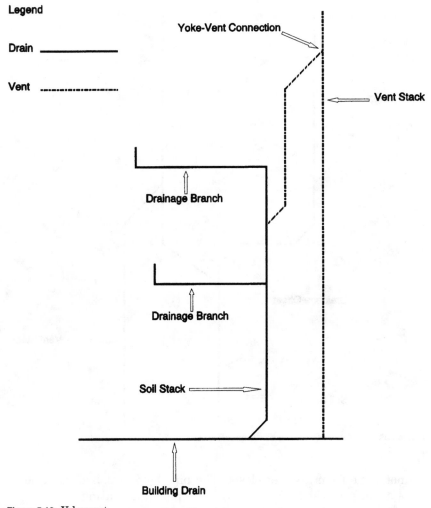

Figure 5.13 Yoke vent.

When a fixture with a fixture rating of 4 or less and a maximum drain size of 2 in is above the battery, every vertical branch must have a continuous vent. If a fixture with a fixture-unit rating exceeding 4 is present, all fixtures in the battery must be individually vented. Circuit-vented batteries may not receive the drainage from fixtures on a higher level.

Circuit vents should rise vertically from the drainage. However, the vent can be taken off the drainage horizontally if the vent is washed by a fixture with a rating of no more than 4 fixture units. The washing

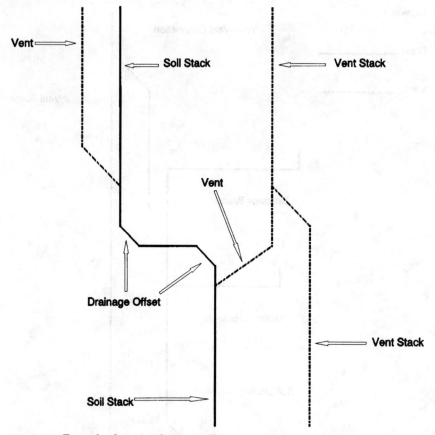

Vent

Soil Stack

Vent Stack

Vent

Drainage Offset

Vent Stack

Soil Stack

Figure 5.14 Example of venting drainage offsets.

cannot come from a water closet. The pipe being washed must be at least as large as the horizontal drainage pipe it is venting.

In zone three, circuit vents may be used to vent up to eight fixtures using a common horizontal drain. Circuit vents must be dry vents, and they should connect to the horizontal drain in front of the last fixture on the branch. The horizontal drain being circuit-vented must not have a grade of more than 1 in/ft. Zone three interprets the horizontal section of drainage being circuit-vented as a vent. If a circuit vent is venting a drain with more than four water closets attached to it, a relief vent must be installed in conjunction with the circuit vent. Figure 5.17 shows you how this would be done.

Vent placement in relation to the trap it serves is important and regulated. The maximum allowable distance between a trap and its vent will depend on the size of the fixture drain and trap. Table 5.11 shows you allowable distances for zone one. Table 5.12 depicts the

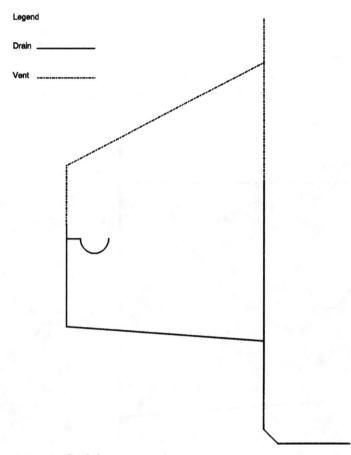

Figure 5.15 Graded-vent connection.

allowable distances for zone two, and Table 5.13 covers the requirements of zone three.

All vents, except those for fixtures with integral traps, should connect above the trap seal. A sanitary-tee fitting should be used when going from a vertical stack vent to a trap. Other fittings, with a longer turn, like a combination wye and eighth bend, will place the trap in more danger of back-siphonage. This goes against the common sense of a smoother flow of water, but the sanitary tee reduces the risk of a vacuum.

Supporting Your Pipe

Vent pipes must be supported. Vents may not be used to support antennas, flag poles, and similar items. Depending upon the type of material

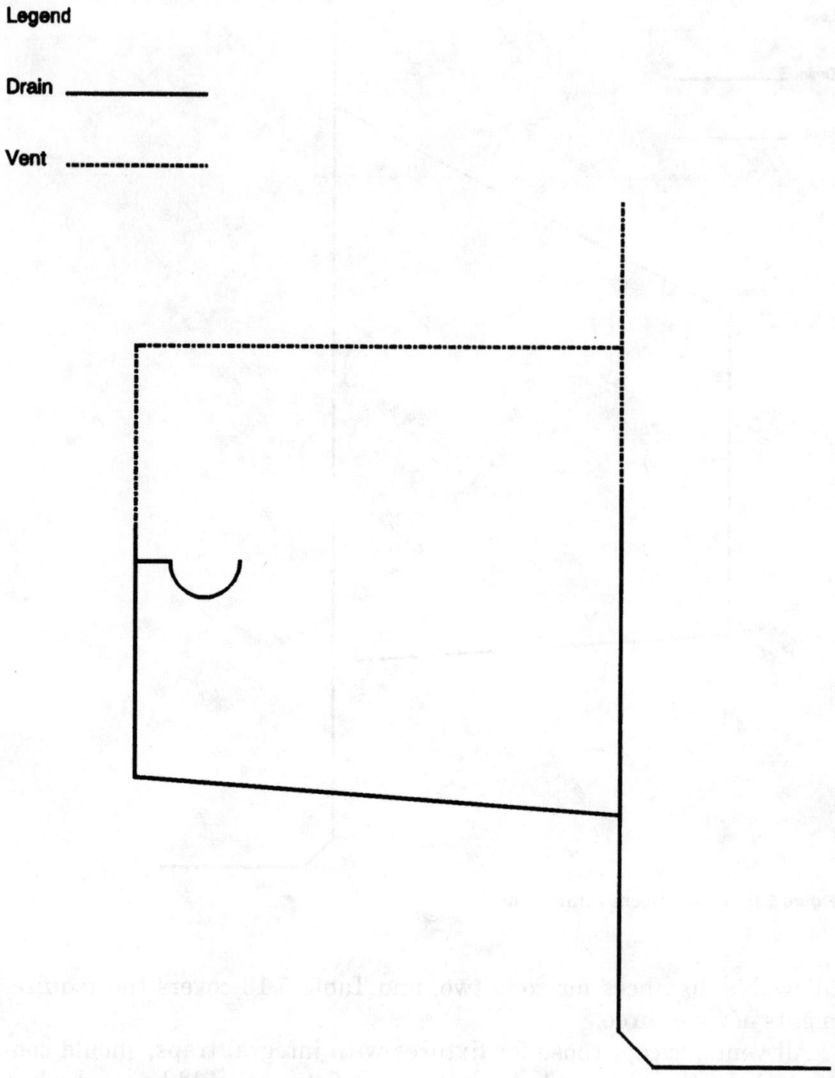

Legend

Drain _____

Vent ....

Figure 5.16 Zone one's level-vent rule.

you are using, and whether the pipe is installed horizontally or vertically, the spacing between hangers will vary. Both horizontal and vertical pipes require support. The regulations in the plumbing code apply
to the maximum distance between hangers. Tables 5.14 and 5.15 give
you the minimums for zone one. Tables 5.16 and 5.17 provide similar
information for zone two. Tables 5.18 and 5.19 give you maximum
hanger intervals for zone three.

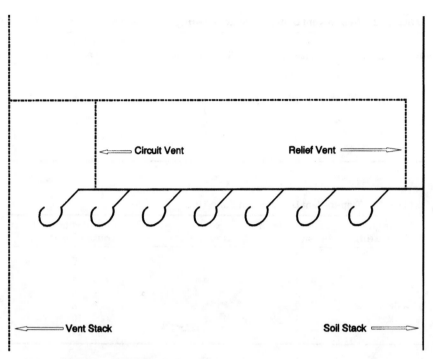

Figure 5.17 Circuit vent with a relief vent.

Some More Venting Regulations
for Zone One

Some interceptors, like those used as a settling tank that discharges through a horizontal indirect waste, are not required to be vented. However, the interceptor receiving the discharge from the unvented interceptor must be properly vented and trapped.

Traps for sinks that are a part of a piece of equipment, like a soda fountain, are not required to be vented when venting is impossible. But these drains must drain through an indirect waste to an approved receptor.

TABLE 5.11   Trap-to-Vent Distances in Zone One

Grade on drain pipe (in)	Size of trap arm (in)	Maximum distance between trap and vent (ft)
¼	1¼	2½
¼	1½	3½
¼	2	5
¼	3	6
¼	4 and larger	10

TABLE 5.12   Trap-to-Vent Distances in Zone Two

Grade on drain pipe (in)	Fixture's drain size (in)	Trap size (in)	Maximum distance between trap and vent (ft)
¼	1¼	1¼	3½
¼	1½	1¼	5
¼	1½	1½	5
¼	2	1½	8
¼	2	2	6
⅛	3	3	10
⅛	4	4	12

TABLE 5.13   Trap-to-Vent Distances in Zone Three

Grade on drain pipe (in)	Fixture's drain size (in)	Trap size (in)	Maximum distance between trap and vent (ft)
¼	1¼	1¼	3½
¼	1½	1¼	5
¼	1½	1½	5
¼	2	1½	8
¼	2	2	6
⅛	3	3	10
⅛	4	4	12

TABLE 5.14   Horizontal Pipe-Support Intervals in Zone One

Support material	Maximum distance between supports (ft)
ABS	4
Cast iron	At each pipe joint*
Galvanized	12
Copper (1½ in and smaller)	6
PVC	4
Copper (2 in and larger)	10

* Cast-iron pipe must be supported at each joint, but supports may not be more than 10 ft apart.

TABLE 5.15   Vertical Pipe-Support Intervals in Zone One*

Type of vent pipe	Maximum distance between supports
Lead pipe	4 ft
Cast iron	At each story
Galvanized	At least every other story
Copper	At each story†
PVC	Not mentioned
ABS	Not mentioned

* All stacks must be supported at their bases.
† Support intervals may not exceed 10 ft.

TABLE 5.16   Horizontal Pipe-Support Intervals in Zone Two

Type of vent pipe	Maximum distance between supports (ft)
ABS	4
Cast iron	At each pipe joint
Galvanized	12
PVC	4
Copper (2 in and larger)	10
Copper (1½ in and smaller)	6

TABLE 5.17   Vertical Pipe-Support Intervals in Zone Two*

Type of vent pipe	Maximum distance between supports (ft)
Lead pipe	4
Cast iron	At each story[†]
Galvanized	At each story[‡]
Copper (1¼ in)	4
Copper (1½ in and larger)	At each story
PVC (1½ in and smaller)	4
PVC (2 in and larger)	At each story
ABS (1½ in and smaller)	4
ABS (2 in and larger)	At each story

* All stacks must be supported at their bases.
[†] Support intervals may not exceed 15 ft.
[‡] Support intervals may not exceed 30 ft.

TABLE 5.18   Horizontal Pipe-Support Intervals in Zone Three

Type of vent pipe	Maximum distance between supports (ft)
Lead pipe	Continuous
Cast iron	5*
Galvanized	12
Copper tube (1¼ in)	6
Copper tube (1½ in and larger)	10
ABS	4
PVC	4
Brass	10
Aluminum	10

* Or at every joint.

TABLE 5.19   Vertical Pipe-Support Intervals in Zone Three

Type of vent pipe	Maximum distance between supports (ft)
Lead pipe	4
Cast iron	15
Galvanized	15
Copper tubing	10
ABS	4
PVC	4
Brass	10
Aluminum	15

Other Venting Requirements for Zone Two

All soil stacks that receive the waste of at least two vented branches must be equipped with a stack vent or a main stack vent. Except when approved, fixture drainage may not be allowed to enter a stack at a point above a vent connection. Side-inlet closet bends are allowed to accept the connection of fixtures that are vented. However, these connections may not be used to vent a bathroom unless the connection is washed by a fixture. All fixtures dumping into a stack below a higher fixture must be vented except when special approval is granted for a variance. Stack vents and vent stacks must connect to a common vent header prior to vent termination.

Traps for sinks that are a part of a piece of equipment, like a soda fountain, are not required to be vented when venting is impossible. But, these drains must be piped in accordance with the combination waste and vent regulations for zone two.

Up to two fixtures, set back to back or side by side, within the allowable distance between the traps and their vents may be connected to a common horizontal branch that is vented by a common vertical vent. However, the horizontal branch must be one pipe size larger than normal. When applying this rule, the following ratings apply: shower drains, 3-in floor drains, 4-in floor drains, pedestal urinals, and water closets with fixture-unit ratings of 4 shall be considered to have 3-in drains.

Some fixture groups are allowed to be stack-vented without individual back vents. These fixture groups must be located in one-story buildings or must be located on the top floor of the building, with some special provisions. Fixtures located on the top floor must connect independently to the soil stack, and the bathing units and water closets must enter the stack at the same level. Table 5.20 shows the fixtures allowed to be vented in this fashion.

This same stack-venting procedure can be adapted to work with fixtures on lower floors. The stack being stack-vented must enter the main soil stack though a vertical eighth bend and wye combination. The drainage must enter above the eighth bend. A 2-in vent must be installed on the fixture group. This vent must be 6 in above the flood-level rim of the highest fixture in the group.

TABLE 5.20   Stack-Venting Without Individual Vents in Zone Two

Fixtures allowed to be stack-vented without individual vents*
Water closets
Basins
Bathtubs
Showers
Kitchen sinks, with or without dishwasher and garbage disposer

* Restrictions apply to this type of installation.

Some fixtures are allowed to be served by a horizontal waste that is within a certain distance of a vent. When piped in this manner, bathtubs and showers are both required to have 2-in P-traps. These drains must run with a minimum grade of ¼-in/ft. A single drinking fountain can be rated as a lavatory for this type of piping. On this type of system, fixture drains for lavatories may not exceed 1¼ in, and sink drains cannot be larger than 1½ in, in diameter.

In multistory situations, it is possible to drain up to three fixtures into a soil stack above the highest water closet or bathtub connection, without reventing. To do this, certain requirements must be met. These requirements are as follows:

- Minimum stack size of 3 in is required.

- Approved fixture-unit load on stack is met.

- All lower fixtures must be properly vented.

- All individually unvented fixtures are within allowable distances of the main vent.

- Fixture openings shall not exceed the size of their traps.

- All code requirements must be met and approved.

Working with a Combination Waste and Vent System

Most jurisdictions limit the extent to which fixtures can be served by a combination waste and vent system, but not all. In many locations it is a code violation to include a toilet on a combination system, but Maine, for example, will allow toilets on a combination waste and vent system. Since combination waste and vent systems can get you into a sticky situation, you should consult your local code officer before using such a system. However, this is how the system works, in general.

The types of fixtures you are allowed to connect to with a combination waste and vent system may be limited. In some areas the only fixtures allowed on the combination system are floor drains, standpipes, sinks, and lavatories. Other areas allow showers, bathtubs, and even toilets to be installed with the combo system. You will have to check your local regulations to see how they affect your choice in types of plumbing systems.

It is intended that the combination waste and vent system will be mainly made up of horizontal piping. Generally, the only vertical piping is the vertical risers to lavatories, sinks, and standpipes. These vertical pipes may not normally exceed 8 ft in length. This type of system relies on an oversized drain pipe to provide air circulation for drainage. The pipe is often required to be twice the size required for a drain vented normally. The combination system typically must have at least one vent. The vent should connect to a horizontal drain pipe.

Any vertical vent must rise to a point at least 6 in above the highest fixture being served before it may take a horizontal turn. In a combination system the pipes are rated for fewer fixture units. A 3-in pipe connecting to a branch or stack may only be allowed to carry 12 fixture units. A 4-in pipe, under the same conditions, could be restricted to conveying 20 fixture units. Similarly, a 2-in pipe might only handle 3 fixture units, and a 1½-inch pipe may not be allowed. The ratings for these pipes can increase when the pipes are connecting to a building drain.

Stack vents are allowed but not always in the normal way. All fixtures on a combo system may be required to enter the stack vent individually, as opposed to on a branch, as would normally be the case. A stack vent used in a combo system generally must be a straight vertical vent, without offsets. The stack vent usually cannot even be offset vertically; it simply cannot be offset. This rule is different in some locations, so check with your local plumbing inspector to see if you are affected by the no-offset rule.

Since stack vents are common, and often required, in a combination system, you must know how to size these pipes. The sizing is generally done based on the number of fixture units entering the stack. Here is an example of how a stack vent for a combo system might be sized in zone three.

Since not all pipes run in conjunction with a combination waste and vent system have to apply to the combo rules, it is possible that you would have a 1½-in pipe entering a stack. The 1½-in pipe could only be used if it had an individual vent. It is also possible that the stack vent would be a 1½-in pipe.

First, let's look at the maximum number of fixture units (fu) allowed on a stack; they are as follows:

- 1½-in stack = 2 fu
- 2-in stack = 4 fu
- 3-in stack = 24 fu
- 4-in stack = 50 fu
- 5-in stack = 75 fu
- 6-in stack = 100 fu

When you are concerned with the size of a drain dumping into the stack, there are only two pipe sizes to contend with. All pipe sizes larger than 2-in may dump an unlimited number of fixture units into the stack. A 1½-in pipe may run one fixture unit into the stack, and a 2-in pipe may deliver two fixture units to the stack. Sizing your stack is as simple as finding your fixture-unit load on the chart in your local code book. Compare your fixture-unit load to the chart, and select a pipe size rated for that load.

Again, remember that combination waste and vent systems vary a great deal in what is, and is not, allowed. To show that contrast, here are how the regulations in Maine differ from the ones already described. In Maine, at the time of this writing, there are regulations that allow a much different style of combination waste and vent plumbing. The Maine regulations allow water closets, showers, and bathtubs to be included on the combo system. These regulations are under consideration for change, but at the moment, they are still in effect.

This alternate form of plumbing is only applicable to buildings with two stories or less. The regulations require that the only fixture located in a basement may be a clothes washer. Not only can a water closet be used on this system, but up to three water closets are allowed to enter a single 3-in stack from a branch. You are also allowed to pipe up to two of these branches, each carrying the discharge from up to three toilets, to the stack at the same location. If a toilet is installed in the building, there must be at least one 3-in vent.

Vertical rises from the building drain are allowed to extend 10 feet. A branch may contain up to eight water closets if vented by a circuit vent. Up to three fixtures, not counting floor drains, that are on the same floor level, or a level not exceeding 30 in apart, may be tied into the building drain at a distance greater than 10 ft. Basically, this means that an entire bathroom group can be piped to the building drain without limitation on the distance.

When it comes to sizing on the Maine system, there are some big differences. With a grade of a ¼-in/ft, a 1½-in pipe can carry up to 5 fixture units; a 2-in pipe is allowed up to 12 fixture units. These numbers are much higher than the ones given earlier. As you can see, there are some major differences to be found in the rulings for combination waste and vent systems. Maine currently operates on a plumbing code that is different from the ones used in zones one, two, and three.

Reminder Notes

Zone one

1. All normal venting is sized using the same ratings and table.

2. Fixtures being vented with a common vent must enter the drainage at the same level.

3. Vent venting is restricted to vertical piping.

4. All plumbing fixtures, except for exceptions, must be vented.

5. Vents terminating closer than 10 ft from a building opening must rise at least 3 ft above the opening.

6. Vents subject to freezing must be no less than 2 in in diameter and must extend at least 10 in above the roof. Any change in pipe size must be made at least 12 in before the pipe penetrates into open air.

7. Side-wall vents are prohibited.

8. Buildings with at least 10 stories above the building drain must be equipped with vent stacks. These vent stacks are required to have yoke vents at every five-story interval.

9. Vents may be installed level, without pitch.

Zone two

1. Wet venting is allowed for multiple bathrooms on different floor levels.

2. Any building equipped with plumbing must be equipped with a main vent.

3. Vents terminating above a roof used for purposes other than weather protection must rise at least 5 ft above the roof.

4. Side-wall vents may not terminate under overhangs.

5. Offsets of more than 45° in soil stacks, when there are at least five stories in the building, must be vented.

6. There are many rules on circuit venting; refer to the text for examples.

Zone three

1. Common vents may serve up to three lavatory traps.

2. Zone three has specific requirements for venting sewer sumps.

3. Any plumbing system receiving the discharge from toilets must be equipped with a main stack for venting.

4. Vents must extend at least 12 in above the roof they penetrate.

5. Vents subject to freezing must have minimum diameters of 3 in. Any change in pipe sizing must occur at least 12 in prior to penetration into open air.

6. Buildings with at least five branch intervals must have vent stacks.

7. In buildings with at least five branch intervals, horizontal offsets in drainage piping must be vented.

8. Circuit vents can serve up to eight fixtures.

9. Circuit vents venting more than four water closets must be equipped with relief vents.

10. Refer to the text for more rules on circuit venting.

6

Traps, Cleanouts, Interceptors, and More

We have covered most of the regulations you will need to know about drains and vents. This chapter will round out your knowledge. Here you will learn about traps. Traps have been mentioned before, and you have learned the importance of vents to trap seals, but here you will learn more about traps.

Cleanouts are a necessary part of the drainage system. This chapter will tell you what types of cleanouts you can use and when and where they must be used. Along with cleanouts, back-water valves will be explained. Grease receptors, or grease traps as they are often called, will be explored. By the end of this chapter you should be prepared to tackle just about any drain waste and vent (DWV) job.

Cleanouts

What are cleanouts, and why are they needed? Cleanouts are a means of access to the interior of drainage pipes. They are needed so that blockages in drains may be cleared. Without cleanouts, it is much more difficult to snake a drain. In general, the more cleanouts you have, the better. Plumbing codes establish minimums for the number of cleanouts required and their placement. Let's look at how these regulations apply to you.

Where Are Cleanouts Required?

There are many places in a plumbing system where cleanouts are required. Let's start with sewers. All sewers must have cleanouts. The distances between these cleanouts vary from region to region. Generally, cleanouts will be required where the building drain meets the building sewer. The cleanouts may be installed inside the foundation or outside, but the cleanout opening must extend upward to the finished floor level or the finished grade outside.

Zone two prefers that the cleanouts at the junction of building drains and sewers be located outside. If the cleanout is installed inside, within zone two, it must extend above the flood level rim of the fixtures served by the horizontal drain. When this is not feasible, allowances may be made. Zone three will waive the requirement for a junction cleanout if there is a cleanout of at least a 3-in diameter within 10 ft of the junction.

Once the sewer is begun, cleanouts should be installed every 100 ft. In zone two, the interval distance is 75 ft for 4-in and larger pipe, and 55 ft for pipe smaller than 4-in. Cleanouts are also required in sewers when the pipe takes a change in direction. In zone three, a cleanout is required every time the sewer turns more than 45°. In zone one, a cleanout is required whenever the change in direction is more than 135°.

The cleanouts installed in a sewer must be accessible. This generally means that a standpipe will rise from the sewer to just below ground level. At that point, a cleanout fitting and plug are installed on the standpipe. This allows the sewer to be snaked out from ground level, with little to no digging required.

For building drains and horizontal branches, the cleanout location will depend upon pipe size, but cleanouts are normally required every 50 ft. For pipes with diameters of 4 in, or less, cleanouts must be installed every 50 ft. Larger drains may have their cleanouts spaced at 100-ft intervals. Cleanouts are also required on these pipes with a change in direction. For zone three, the degree of change is anything in excess of 45°. Cleanouts must be installed at the end of all horizontal drain runs. Zone one does not require cleanouts at 50-ft intervals, only at 100-ft intervals.

As with most rules, there are some exceptions to these. Zone one offers some exceptions to the cleanout requirements for horizontal drains. The following exceptions apply only to zone one. If a drain is less than 5 ft long and is not draining sinks or urinals, a cleanout is not required. A change in direction from a vertical drain with a fifth bend does not require a cleanout. Cleanouts are not required on pipes, other than building drains and their horizontal branches, that are above the first-floor level.

P-traps and water closets are often allowed to act as cleanouts. When these devices are approved for cleanout purposes, the normally required cleanout fitting and plug at the end of a horizontal pipe run may be eliminated. Not all jurisdictions will accept P-traps (Fig. 6.1) and toilets as cleanouts; check your local requirements before omitting standard cleanouts.

Cleanouts must be installed in a way that the cleanout opening is accessible and allows adequate room for drain cleaning. The cleanout must be installed to go with the flow. This means that when the cleanout plug is removed, a drain-cleaning device should be able to enter the fitting and the flow of the drainage pipe without trouble.

When you are installing your plumbing in zone three, you must include a cleanout at the base of every stack (Fig. 6.2). This is good procedure at any time, but it is not required by all codes. The height of this cleanout should not exceed 4 ft. Many plumbers install test tees at these locations to plug their stacks for pressure testing (Fig. 6.3). The test tee doubles as a cleanout.

When the pipes holding cleanouts will be concealed, the cleanout must be made accessible. For example, if a stack will be concealed by a finished wall, provisions must be made for access to the cleanout. This

Figure 6.1 P-trap.

Figure 6.2 Test tee.

Figure 6.3 Test tee with test ball installed.

access could take the form of an access door, or the cleanout could simply extend past the finished wall covering. If the cleanout is serving a pipe concealed by a floor, the cleanout must be brought up to floor level and made accessible. This ruling applies not only to cleanouts installed beneath concrete floors but also to cleanouts installed in crawl spaces, with very little room to work.

What Else Do I Need to Know about Cleanouts?

There is still more to learn about cleanouts. Size is one of the lessons to be learned. Cleanouts are required to be the same size as the pipe they are serving, unless the pipe is larger than 4 in. If you are installing a 2-in pipe, you must install 2-in cleanouts. However, when a P-trap is allowed for a cleanout, it may be smaller than the drain (Fig. 6.4). An example would be a 1¼-in trap on a 1½-in drain. Remember though, not all code enforcement officers will allow P-traps as cleanouts, and they may require the P-trap to be the same size as the drain if the trap is allowed as a cleanout. Once the pipe size exceeds 4 in, the cleanouts used should have a minimum size of 4 in.

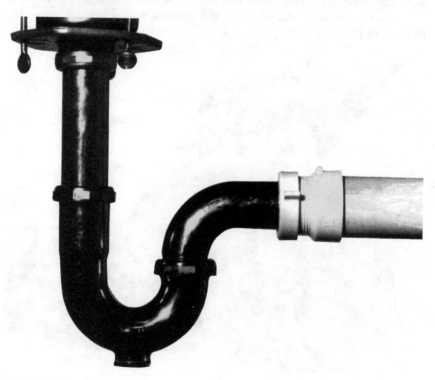

Figure 6.4 Slip-nut P-trap.

When cleanouts are installed, they must provide adequate clearance for drain cleaning (Fig. 6.5). The clearance required for pipes with diameters of 3 in, or more is 18 in. Smaller pipes require a minimum clearance of 12 in in front of their cleanouts. Many plumbers fail to remember this regulation. It is common to find cleanouts pointing toward floor joists or too close to walls. You will save yourself time and money by committing these clearance distances to memory.

Zone one takes the clearance rules a step further. In zone one, when a cleanout is installed in a floor, it must have a minimum height clearance of 18 in and a minimum horizontal clearance of 30 in. No under-floor cleanout is allowed to be placed more than 20 ft from an access opening.

Acceptable Types of Cleanouts

Cleanout plugs and plates must be easily removed. Access to the interior of the pipe should be available without undue effort or time. Cleanouts can take on many appearances (Fig. 6.6). The U bend of a P-trap can be considered a cleanout, depending upon local interpretation. A rubber cap, held onto the pipe by a stainless steel clamp (Fig. 6.7), can serve as a cleanout. The standard female adapter and plug are a fine cleanout. Test tees will work as cleanouts. Special cleanouts, designed to allow rodding of a drain in either direction, are acceptable.

Figure 6.5 Raised-head cleanout plug.

Figure 6.6 Recessed-head cleanout plug.

Very Big Cleanouts

The ultimate cleanout is a manhole. You can think of manholes as very big cleanouts. When a pipe's diameter exceeds 10 in in zone three, or 8 in in zone two, manholes replace cleanouts. Manholes are required every 400 ft in zone three and every 300 ft in zones one and two. In addition, they are required at all changes in direction, elevation, grade,

Figure 6.7 Rubber cap.

and size. Manholes shall be protected against flooding and must be equipped with covers to prevent the escape of gases. Zone one requires connections with manholes to be made with flexible compression joints. These connections must not be made closer than 1 ft to the manhole and not farther than 3 ft away from it.

Traps

Traps are required on drainage-type plumbing fixtures. With some fixtures, like toilets, traps are not apparent because they are an integral part of the fixture. The following regulations do not apply to integral traps, which being a part of a fixture, are governed by regulations controlling the use of approved fixtures. We have already talked about trap seals, so now let's learn more about traps.

P-traps

P-traps are the traps most frequently used in modern plumbing systems. These traps are self-cleaning and frequently have removable U bends that may act as cleanouts, pending local approval. P-traps must be properly vented. Without adequate venting, the trap seal can be removed by back pressure.

S-traps

S-traps were very common when most plumbing drains came up through the floor instead of out of a wall. Many S-traps are still in operation, but they are no longer allowed in new installations. S-traps can lose their trap seal through self-siphoning.

Drum traps

Drum traps are not normally allowed in new installations without special permission from the code officer. The only occasion when drum traps are still used frequently is when they are installed with a combination waste and vent system.

Bell traps

Bell traps are not allowed for use in new installations.

House traps

House traps are no longer allowed; they represent a double trapping of all fixtures. House traps were once installed where the building drain joined with the sewer. Most house traps were installed inside the structure, but a fair number were installed outside, underground. Their purpose was to prevent sewer gas from coming out of the sewer and into the plumbing system, but house traps make drain cleaning very difficult and they create a double-trapping situation, which is not allowed. This regulation, like most regulations, is subject to amendment and variance by the local code official.

Crown-vented traps

Crown-vented traps are not allowed in new installations. These traps have a vent rising from the top of the trap. As you learned earlier, crown venting must be done at the trap arm, not the trap.

Other traps

Traps that depend on moving parts or interior partitions are not allowed in new installations.

Does Every Fixture Require an Individual Trap?

Basically, every fixture requires an individual trap, but there are exceptions. One such exception is the use of a continuous waste to connect the drains from multiple sink bowls to a common trap. This is done frequently with kitchen sinks.

There are some restrictions involving the use of continuous wastes. Let's take a kitchen sink as an example. When you have a double-bowl sink, it is okay to use a continuous waste, as long as the drains from each bowl are no more than 30 in apart and neither bowl is more than 6 in deeper than the other bowl. Zone one requires that all sinks connected to a continuous waste must be of equal depth. Exceptions to this rule do exist.

What if your sink has three bowls? Three-compartment sinks may be connected with a continuous waste. You may use a single trap to collect the drainage from up to three separate sinks or lavatories as long as they are next to each other and in the same room. But, the trap must be in a location central to all sinks or lavatories.

Trap Sizes

Trap sizes are determined by the local code. Tables 6.1, 6.2, and 6.3 give you examples of commonly accepted trap sizes. A trap may not be larger than the drain pipe it discharges into.

TABLE 6.1   Recommended Trap Sizes for Zone One

Type of fixture	Trap size (in)
Bathtub	1½
Shower	2
Residential toilet	Integral
Lavatory	1¼
Bidet	1½
Laundry tub	1½
Washing machine standpipe	2
Floor drain	2
Kitchen sink	1½
Dishwasher	1½
Drinking fountain	1¼
Public toilet	Integral

TABLE 6.2   Recommended Trap Sizes for Zone Two

Type of fixture	Trap size (in)
Bathtub	1½
Shower	2
Residential toilet	Integral
Lavatory	1¼
Bidet	1½
Laundry tub	1½
Washing machine standpipe	2
Floor drain	2
Kitchen sink	1½
Dishwasher	1½
Drinking fountain	1
Public toilet	Integral

TABLE 6.3   Recommended Trap Sizes for Zone Three

Type of fixture	Trap size (in)
Bathtub	1½
Shower	2
Residential toilet	Integral
Lavatory	1¼
Bidet	1¼
Laundry tub	1½
Washing machine standpipe	2
Floor drain	2
Kitchen sink	1½
Dishwasher	1½
Drinking fountain	1¼
Public toilet	Integral
Urinal	2

Tailpiece Length

The tailpiece between a fixture drain and the fixture's trap may not exceed 24 in.

Standpipe Height

A standpipe, when installed in zone three, must extend at least 18 in above its trap but may not extend more than 30 in above the trap. Zone two prohibits the standpipe from extending more than 4 ft from the trap. Zone one requires the standpipe not to exceed a height of more than 2 ft above the trap. Plumbers installing laundry standpipes often forget this regulation. When setting your fitting height in the drainage pipe, keep in mind the height limitations on your standpipe. Otherwise, your take-off fitting may be too low, or too high, to allow your standpipe receptor to be placed at the desired height. Traps for kitchen sinks may not receive the discharge from a laundry tub or clothes washer.

Proper Trap Installation

There is more to proper trap installation than location and trap selection. Traps must be installed level in order for the trap seal to function properly. An average trap seal will consist of 2 in of water. Some large traps may have a seal of 4 in, and where evaporation is a problem, deep-sealing traps may have a deeper water seal. The positioning of the trap is critical for the proper seal. If the trap is cocked, the water seal will not be uniform and may contribute to self-siphoning.

When a trap is installed below grade and must be connected from above grade, the trap must be housed in a box of a sorts (Fig. 6.8). An example of such a situation would be a trap for a tub waste. When installing a bathtub on a concrete floor, the trap is located below the floor. Since the trap cannot be reasonably installed until after the floor is poured, access must be made for the connection. This access, frequently called a tub box or trap box, must provide protection against water, insect, and rodent infiltration.

When Is a Trap Not a Trap?

One type of trap we have not yet discussed is a grease trap. The reason we haven't talked about grease traps is that grease traps are not really traps; they are interceptors. They are frequently called grease traps, but they are actually grease interceptors. There is a big difference between a trap and an interceptor.

Figure 6.8 Trap box.

Traps are meant to prevent sewer gas from entering a building. Traps do not restrict what goes down the drain, only what comes up the drain. Of course, traps do prevent objects larger than the trap from entering the drain, but this is not their primary objective.

Interceptors, on the other hand, are designed to control what goes down a drain. Interceptors are used to keep harmful substances from entering the sanitary drainage systems. Separators, because they separate the materials entering them and retain certain materials, while allowing others to continue into the drainage system, are also required in some circumstances. Interceptors are used to control grease, sand, oil, and other materials.

Interceptors and separators are required when conditions provide opportunity for harmful or unwanted materials to enter a sanitary drainage system. For example, a restaurant is required to be equipped with a grease interceptor because of the large amount of grease present in commercial food establishments. An oil separator would be required for a building where automotive repairs are made. Interceptors and separators must be designed for each individual situation. There is no rule-of-thumb method of choosing the proper interceptor or separator without expert design.

There are some guidelines provided in plumbing codes for interceptors and separators. The capacity of a grease interceptor is based on two factors: grease retention and flow rate. These determinations are typically made by a professional designer. The size of a receptor or separator is also normally determined by a design expert.

Interceptors for sand and other heavy solids must be readily accessible for cleaning. These units must contain a water seal of not less than 6 in, except in zone two, where the minimum water depth is 2 in. When an interceptor is used in a laundry, a water seal is not required. Laundry receptors, used to catch lint, string, and other objects, are usually made of wire, and they must be easily removed for cleaning. Their purpose is to prevent solids with a diameter of ½ in, or more, from entering the drainage system.

Other types of separators are used for various plants, factories, and processing sites. The purpose of all separators is to keep unwanted objects and substances from entering the drainage system. Vents are required if it is suspected that these devices will be subject to the loss of a trap seal. All interceptors and separators must be readily accessible for cleaning, maintenance, and repairs.

Back-Water Valves

Back-water valves are essentially check valves. They are installed in drains and sewers to prevent the backing-up of waste and water in the drain or sewer. Back-water valves are required to be readily accessible and installed whenever a drainage system is likely to encounter backups from the sewer.

The intent behind back-water valves is to prevent sewers from backing up into individual drainage systems. Buildings that have plumbing fixtures below the level of the street, where a main sewer is installed, are candidates for back-water valves.

This concludes our section on traps, cleanouts, interceptors, and other drainage-related regulations. While this is a short chapter, it is an important one. You may not have a need for installing manholes or back-water valves every day, but, as a plumber, you will frequently work with traps and cleanouts.

Reminder Notes

Zone one

1. Cleanouts are required in sewers when the pipe makes a change in direction of more than 135°.

2. Cleanouts in all horizontal drains may be spaced up to 100 ft apart, rather than 50 ft, as in zones two and three.

3. Cleanouts are not required in horizontal drains that are less than 5 ft in length, unless the pipe is draining urinals or sinks.

4. A change in direction from a vertical drain with a fifth bend does not require a cleanout.

5. Cleanouts are not required on pipes, other than building drains and their horizontal branches, that are above the first-floor level.

6. Cleanouts extending from a floor must have a vertical clearance of 18 in and a horizontal clearance of 30 in. No underfloor cleanout may be placed more than 20 ft from an access opening.

7. Manhole connections must be made with flexible connections. These connections must be made at least 1 ft from the manhole but not farther than 3 ft from it.

8. When multiple sinks are connected to a single trap, with a continuous waste, all sink bowls must be of equal depth.

9. Laundry standpipes may not extend more than 2 ft above the traps.

10. Tubular traps must have a minimum rating of 17 gauge; however, these thin-walled traps are not allowed for use with urinals.

Zone two

1. Cleanouts at the junction of a building drain and sewer should be located outside. If they are located inside, the cleanout should extend above the flood-level rim of the fixtures being served by the horizontal pipe. If getting the cleanout above the flood-level rim is not feasible, allowances may be made.

2. Cleanouts in a 4-in, or larger, sewer should be installed with intervals of not more than 75 ft between them. If the sewer pipe is smaller than 4 in, the interval should be reduced to 50 ft.

3. Manholes are required on sewers with diameters larger than 8 in.

4. Laundry standpipes may not extend more than 4 ft above the trap.

5. Tubular traps must have a minimum rating of 20 gauge.

6. Sand interceptors must have a minimum water seal of 2 in.

Zone three

1. Cleanouts at the junction of a building drain and building sewer may be eliminated when a cleanout with a minimum diameter of 3 in is within 10 ft of the junction.

2. Cleanouts are required in the sewer when the pipe takes a change in direction of more than 45°.

3. Cleanouts are required on all horizontal drains when a change in direction of more than 45° is needed.

4. A cleanout is required at the base of each drainage stack.

5. Manholes are required on sewers with diameters larger than 10 in.

6. Manholes must not be more than 400 ft apart.

7. Laundry standpipes must extend at least 18 in above the trap, but not more than 30 in.

Fixtures

There is much more to fixtures than meets the eye. Fixtures are a part of the final phase of plumbing. When you are planning a plumbing system, you must know which fixtures are required, what type of fixtures are needed, and how they must be installed. This chapter will guide you through the myriad of fixtures available and how they may be used.

What Fixtures Are Required?

The number and types of fixtures required will depend on local regulations and the use of the building in which they are being installed. Your code book will provide you with information on what is required and how many of each type of fixture is needed. These requirements are based on the use of the building housing the fixtures and the number of people that may be using the building. Here are some examples.

Single-family residence

When you are planning fixtures for a single-family residence, you must include certain fixtures. If you choose to install more than the minimum, that's fine, but you must install the minimum number of required fixtures, which are as follows:

- One toilet
- One lavatory
- One bathing unit
- One kitchen sink
- One hook-up for a clothes washer

Multifamily buildings

The minimum requirements for a multifamily building are the same as those for a single-family dwelling, but the requirements are that each dwelling in the multifamily building must be equipped with the minimum fixtures. There is one exception—the laundry hook-up. With a multifamily building, laundry hook-ups are not required in each dwelling unit. In zone three, it is required that a laundry hook-up be installed for common use when the number of dwelling units is 20. For each interval of 20 units, you must install a laundry hook-up. For example, in a building with 40 apartments, you would have to provide two laundry hook-ups. If the building had 60 units, you would need three hook-ups. In zone one, the dwelling-unit interval is 10 rental units. Zone two requires one hook-up for every 12 rental units but no less than two hook-ups for buildings with at least 15 units.

Nightclubs

When you get into businesses and places of public assembly, like nightclubs, the ratings are based on the number of people likely to use the facilities. In a nightclub, the minimum requirements for zone three are as follows:

- Toilets—one toilet for every 40 people
- Lavatories—one lavatory for every 75 people
- Service sinks—one service sink
- Drinking fountains—one drinking fountain for every 500 people
- Bathing units—none

Day-care facilities

The minimum number of fixtures for a day-care facility in zone three are as listed below:

- Toilets—one toilet for every 15 people
- Lavatories—one lavatory for every 15 people
- Bathing units—one bathing unit for every 15 people
- Service sinks—one service sink
- Drinking fountains—one drinking fountain for every 100 people

In contrast, zone two only requires the installation of toilets and lavatories in day-care facilities. The ratings for these two fixtures are the same as in zone three, but the other fixtures required by zone three are not required in zone two. This type of rating system will be found in your local code and will cover all the normal types of building uses.

In many cases, facilities will have to be provided in separate bathrooms, to accommodate each sex. When installing separate bathroom facilities, the number of required fixtures will be divided equally between the two sexes, unless there is cause and approval for a different appropriation.

Some types of buildings do not require separate facilities. For example, zone three does not require the following buildings to have separate facilities: residential properties and small businesses where less than 15 employees work or where less than 15 people are allowed in the building at the same time.

Zone two does not require separate facilities in the following buildings: offices with less than 1200 ft^2, retail stores with less than 1500 ft^2, restaurants with less than 500 ft^2, self-serve laundries with less than 1400 ft^2, and hair salons with less than 900 ft^2.

Employee and customer facilities

There are some special regulations pertaining to employee and customer facilities. For employees, toilet facilities must be available to employees within a reasonable distance and with relative ease of access. For example, zone three requires these facilities to be in the immediate work area; the distance an employee is required to walk to the facilities may not exceed 500 ft. The facilities must be located in a manner so that employees do not have to negotiate more than one set of stairs for access to the facilities. There are some exceptions to these regulations, but in general, these are the rules.

It is expected that customers of restaurants, stores, and places of public assembly shall have toilet facilities. In zone three, this is based on buildings capable of holding 150 or more people. Buildings in zone three with an occupancy rating of less than 150 people are not required to provide toilet facilities, unless the building serves food or beverages. When facilities are required, they may be placed in individual buildings or in a shopping mall situation, in a common area, not more than 500 ft from any store or tenant space. These central toilets must be placed so that customers will not have to use more than one set of stairs to reach them.

Zone two uses a square-footage method to determine minimum requirements in public places. For example, retail stores are rated as having an occupancy load of one person for every 200 ft^2 of floor space. This type of facility is required to have separate facilities when the store's square footage exceeds 1500 ft^2. A minimum of one toilet is required for each facility when the occupancy load is up to 35 people.

One lavatory is required in each facility, for up to 15 people. A drinking fountain is required for occupancy loads up to 100 people.

Handicap Fixtures

Handicap fixtures are not cheap; you cannot afford to overlook them when bidding a job. The plumbing code will normally require specific minimum requirements for handicap-accessible fixtures in certain circumstances. It is your responsibility to know when handicap facilities are required. There are also special regulations pertaining to how handicap fixtures shall be installed. We are about to embark on a journey into handicap fixtures and their requirements.

When you are dealing with handicap plumbing, you must mix the local plumbing code with the local building code. These two codes work together in establishing the minimum requirements for handicap plumbing facilities. When you step into the field of handicap plumbing, you must play by a different set of rules. Handicap plumbing is like a different code of its own.

Where are handicap fixtures required?

Most buildings frequented by the public are required to have handicap-accessible plumbing fixtures. The following handicap examples are based on zone three requirements. Zones one and two do not go into as much detail on handicap requirements in their plumbing codes.

Single-family homes and most residential multifamily dwellings are exempt from handicap requirements. A rule-of-thumb standard for most public buildings is the inclusion of one toilet and one lavatory for handicap use.

Hotels, motels, inns, and the like are required to provide a toilet (Fig. 7.1), lavatory (Fig. 7.2), bathing units (Fig. 7.3), and kitchen sink, where applicable, for handicap use. Drinking fountains may also be required. This provision will depend on the local plumbing and building codes. If plumbing a gang shower arrangement, such as in a school gym, at least one of the shower units must be handicap-accessible. Door sizes and other building code requirements must be observed when dealing with handicap facilities. There are local exceptions to these rules; check with your local code officers for current, local regulations.

Installation considerations

When it comes to installing handicap plumbing facilities, you must pay attention to the plumbing and building codes. In most cases, approved blueprints will indicate the requirements of your job, but in rural

4094 Atlas Elongated Rim
- 18″ rim height handicapped.
- 12″ rough-in.
- Anti-Siphon ballock.
- 3.5 G.P.F.

Figure 7.1 Handicap toilet. (*Courtesy Universal-Rundle Corporation*)

areas, you may not enjoy the benefit of highly detailed plans and specifications. When it comes time for a final inspection, the plumbing must pass muster along with the open space around the fixtures. If the inspection is failed, your pay is held up and you are likely to incur unexpected costs. This section will apprise you of what you may need to know. It is not all plumbing, but it is all needed information when working with handicap facilities.

Handicap toilet facilities

When you think of installing a handicap toilet, you probably think of a toilet that sits high off the floor. But, do you think of the grab bars and partition dimensions required around the toilet? Some plumbers don't,

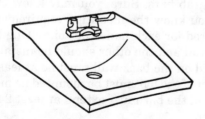

4682 8″cc/4683 4″cc — 27″ x 20″ Wheelchair

Figure 7.2 Handicap lavatory. (*Courtesy Universal-Rundle Corporation*)

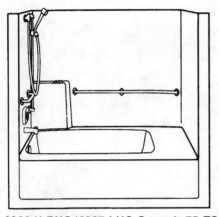

6266-H RHO/6267 LHO Summit 75 TS
- Molded-in seat.
- One-piece seamless construction.
- 1½" diameter safety grab bars.
- Slip resistant bottom.

Figure 7.3 Handicap bathtub. (*Courtesy Universal-Rundle Corporation*)

but they should. The door to a privacy stall for a handicap toilet must provide a minimum of 32 in of clear space.

The distance between the front of the toilet and the closed door must be at least 48 in. It is mandatory that the door open outward, away from the toilet. Think about it; how could a person in a wheelchair close the door if the door opened into the toilet? These facts may not seem like your problem, but if your inspection doesn't pass, you don't get paid.

The width of a water closet compartment for handicap toilets must be a minimum of 5 ft. The length of the privacy stall shall be at least 56 in for wall-mounted toilets and 59 in for floor-mounted models. Unlike regular toilets that require a rough-in of 15 in to the center of the drain from a side wall, handicap toilets require the rough-in to be at least 18 in from the side wall.

Then, there are the required grab bars. Sure, you may know that grab bars are required, but do you know the mounting requirements for the bars? Two bars are required for each handicap toilet. One bar should be mounted on the back wall and the other should be installed on the side wall. The bar mounted on the back wall must be at least 3 ft long. The first mounting bracket of the bar must be mounted no more than 6 in from the side wall. Then, the bar must extend at least 24 in past the center of the toilet's drain.

The bar mounted on the side wall must be at least 42 in long. The bar should be mounted level and with the first mounting bracket located no more than 1 ft from the back wall. The bar must be mounted on the side wall that is closest to the toilet. This bar must extend to a point at least 54 in from the back wall. If you do your math, you will see that a 42-in bar is pushing the limits on both ends. A longer bar will allow more assurance of meeting the minimum requirements.

When a lavatory will be installed in the same toilet compartment, the lavatory must be installed on the back wall in a way that its closest point to the toilet is no less than 18 in from the center of the toilet's drain. When a privacy stall of this size and design is not suitable, there is an option. Another way to size the compartment to house a handicap toilet and lavatory is available. There may be times when space restraints will not allow a stall with a width of 5 ft. In these cases, you may position the fixture differently and use a stall with a width of only 3 ft. In these situations, the width of the privacy stall may not exceed 4 ft.

The depth of the compartment must be at least 66 in when wall-mounted toilets are used. The depth extends to a minimum of 69 in with the use of a floor-mounted water closet. The toilet requires a minimum distance from side walls of 18 in to the center of the toilet drain. If the compartment is more than 3 ft wide, grab bars are required, with the same installation methods as described before.

It the stall is made at the minimum width of 3 ft, grab bars, with a minimum length of 42 in, are required on each side of the toilet. These bars must be mounted no more than 1 ft from the back wall, and they must extend a minimum of 54 in from the back wall. If a privacy stall is not used, the side wall clearances and the grab bar requirements are the same as listed in these two examples. To determine which set of rules to use, you must assess the shape of the room when no stall is present.

If the room is laid out in a fashion like the first example, use the guidelines for grab bars as listed there. If, on the other hand, the room tends to meet the description of the last example, use the specifications in that example. In both cases, the door to the room may not swing into toilet area.

Handicap fixture design

Handicap fixtures are specially designed for people with less physical ability than the general public. The differences in handicap fixtures may appear subtle, but they are important. Let's look at the requirements a fixture must meet to be considered a handicap fixture.

Toilets. Toilets will have a normal appearance, but they will sit higher above the floor than a standard toilet. A handicap toilet will rise to a height of between 16 and 20 in off the finished floor; 18 in is a common height for most handicap toilets. There are many choices in toilet style; they include the following:

- Siphon jet
- Siphon wash
- Siphon vortex
- Reverse trap
- Blowout

Sinks and lavatories.   Visually, handicap sinks, lavatories, and faucets may appear to be standard fixtures, but their method of installation is regulated and the faucets are often unlike a standard faucet. Handicap sinks and lavatories must be positioned to allow a person in a wheelchair to use them easily.

The clearance requirements for a lavatory are numerous. There must be at least 30 in of clearance in front of the lavatory. This clearance must extend 30 in from the front edge of the lavatory or countertop, whichever protrudes the farthest, and to the sides. If you can sit a square box, with a 30- by 30-in dimension, in front of the lavatory or countertop, you have adequate clearance for the first requirement. This applies to kitchen sinks and lavatories.

The next requirement calls for the top of the lavatory to be no more than 35 in from the finished floor. For a kitchen sink, the maximum height is 34 in. Then, there is knee clearance to consider. The minimum allowable knee clearance requires 29 in in height and 8 in in depth. This is measured from the face of the fixture, lavatory, or kitchen sink. Toe clearance is another issue. A space 9 in high and 9 in deep is required, as a minimum, for toe space. The last requirement deals with hot-water pipes. Any exposed hot-water pipes must be insulated or shielded to prevent users of the fixture from being burned.

Sink and lavatory faucets.   Handicap faucets frequently have blade handles. The faucets must be located no more than 25 in from the front edge of the lavatory or counter, whichever is closest to the user. The faucets could use wing handles, single-handles, or push buttons to be operated, but the operational force required by the user shall not be more than 5 lb.

Bathing units.   Handicap bathtubs and showers must meet the requirements of approved fixtures, like any other fixture, but they are also required to have special features and installation methods. The special features are required under the code for approved handicap fixtures. The clear space in front of a bathing unit is required to be a minimum of 1440 in^2. This is achieved by leaving an open space of 30 in in front of the unit and 48 in to the sides. If the bathing unit is not accessible from the side, the minimum clearance is increased to an area with a dimension of 48 by 48 in.

Handicap bathtubs are required to be installed with seats and grab bars. A grab bar, for handicap use, must have a diameter of at least 1¼ in. The diameter may not exceed 1½ in. All handicap grab bars are meant to be installed 1½ in from walls. The design and strength of these bars are set forth in the building codes.

The seat may be an integral part of the bathtub, or it may be a removable, after-market seat. The grab bars must be at least 2 ft long. Two of these grab bars are to be mounted on the back wall, one above the other. The bars are to run horizontally. The lowest grab bar must be mounted 9 in above the flood-level rim of the tub. The top grab bar must be mounted a minimum of 33 in, but no more than 36 in, above the finished floor. The grab bars should be mounted near the seat of the bathing unit.

Additional grab bars are required at each end of the tub. These bars should be mounted horizontally and at the same height as the highest grab bar on the back wall. The bar over the faucet must be at least 2 ft long. The bar on the other end of the tub may be as short as 1 ft.

The faucets in these bathing units must be located below the grab bars. The faucets used with a handicap bathtub must be able to operate with a maximum force of 5 lb. A personal, hand-held shower is required in all handicap bathtubs. The hose for the hand-held shower must be at least 5 ft long.

Two types of showers are normally used for handicap purposes. The first type allows the user to leave a wheelchair and shower while sitting on a seat (Fig. 7.4). The other style of shower stall is meant for the user to roll a wheelchair into the stall and shower while seated in the wheelchair (Fig. 7.5).

If the shower is intended to be used with a shower seat, its dimensions should form a square, with 3 ft of clearance. The seat should be no more than 16 in wide and mounted along the side wall. This seat should run the full length of the shower. The height of the seat should be between 17 and 19 in above the finished floor. There should be two grab bars installed in the shower. These bars should be located between 33 and 36 in above the finished floor. The bars are intended to be mounted in an L shape. One bar should be 36 in long and run the length of the seat, mounted horizontally. The other bar should be installed on the side wall of the shower. This bar should be at least 18 in long.

The faucet for this type of shower must be mounted on the wall across from the seat. The faucet must be at least 38 in but not more than 48 in above the finished floor. There must be a hand-held shower installed in the shower. The hand-held shower can be in addition to a

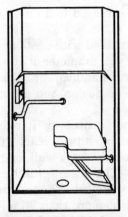

6066-H RHO/6067 LHO Summit 36S
- One-piece seamless construction.
- Fold-down bench.
- 1½" diameter safety grab bars
- Meets ANSI standard A117.1-80.
- Slip resistant floor.

Figure 7.4 Handicap shower with seat. (*Courtesy Universal-Rundle Corporation*)

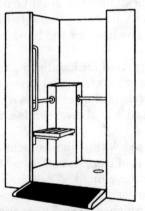

6950 RH Seat/6951 LH Seat Liberte
- Has fold-down seat. Placed at 18" height for easy transfer from wheelchair to seat.
- Two built-in soap shelves.
- One vertical and three horizontal grab bars.
- Inside diameter of 5' for easy wheelchair turn inside stall.
- Entry ramp 36" wide with gentle 8.3% grade.
- Lipped door ledge to prevent rolling out of stall.
- Anti-skid floor mat included.
- White.
- Optional dome (6951) available.

Figure 7.5 Handicap shower with seat and ramp. (*Courtesy Universal-Rundle Corporation*)

fixed shower head, but there must be a hand-held shower on a hose at least 5 ft long installed. The faucet must be able to operate with a maximum force of 5 lb.

Drinking units. The distribution of water from a water cooler or drinking fountain must occur at a maximum height of 36 in above the finished floor. The outlet for drinking water must be located at the front of the unit and the water must flow upward for a minimum distance of 4 in. Levers or buttons to control the operation of the drinking unit may be mounted on the front of the unit or on the side, near the front.

Clearance requirements call for an open space of 30 in in front of the unit and 48 in to the sides. Knee and toe clearances are the same as required for sinks and lavatories. If the unit is made so that the drinking spout extends beyond the basic body of the unit, the width clearance may be reduced from 48 in to 30 in so long as knee and toe requirements are met.

Standard Fixture Installation Regulations

Standard fixtures must also be installed according to local code regulations. There are space limitations, clearance requirements, and predetermined, approved methods for installing standard plumbing fixtures. First, let's look at the space and clearance requirements for common fixtures (Fig. 7.6).

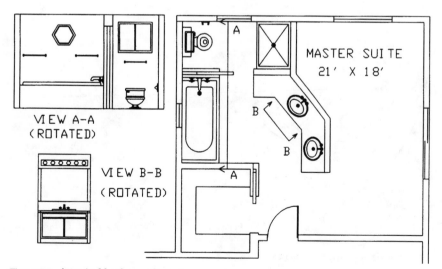

Figure 7.6 A typical bathroom layout.

Standard fixture placement

Toilets and bidets require a minimum distance of 15 in from the center of the fixture's drain to the nearest side wall. These fixtures must have at least 15 in of clear space between the center of their drains and any obstruction, such as a wall, cabinet, or other fixture. With this rule in mind, a toilet or bidet (Fig. 7.7) must be centered in a space of at least 30 in. Figure 7.8 illustrates this placement. Zone one further requires that there be a minimum of 18 in of clear space in front of these fixtures (Fig. 7.9) and that when toilets are placed in privacy stalls, the stalls must be at least 30 in wide and 60 in deep.

Zones one and two require urinals (Fig. 7.10) to be installed with a minimum clear distance of 12 in from the center of their drains to the nearest obstacle on either side. When urinals are installed side by side in zones one and two, the distance between the centers of their drains must be at least 24 in. Zone three requires urinals to have minimum side-wall clearances of at least 15 in. In zone three, the center-to-center distance is a minimum of 30 in. Urinals in zone three must also have a minimum clearance of 18 in in front of them.

Standard fixtures, as with all fixtures, must be installed level and with good workmanship. The fixture should normally be set with an equal distance from walls to avoid a crooked or cocked installation. See Figs. 7.11 and 7.12 for examples of the right and wrong ways to position a toilet. All fixtures should be designed and installed with proper cleaning in mind.

4944 Hygiene II Bidet

Figure 7.7 Bidet. (*Courtesy Universal-Rundle Corporation*)

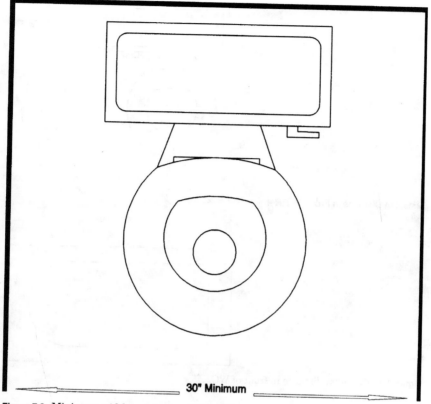

30" Minimum

Figure 7.8 Minimum width requirements for WC.

Bathtubs, showers, vanities, and lavatories should be placed in a manner to avoid violating the clearance requirements for toilets, urinals, and bidets. See Fig. 7.13 for an example of a legal bathroom layout in zone three. Figure 7.14 shows an illegal bathroom grouping.

Securing and sealing fixtures

Some fixtures hang on walls (Fig. 7.15), and others sit on floors (Fig. 7.16). When securing fixtures to walls and floors, there are some rules you must obey. Floor-mounted fixtures, like most residential toilets, should be secured to the floor with the use of a closet flange. The flange is first screwed or bolted to the floor. A wax seal is then placed on the flange, and closet bolts are placed in slots on both sides of the flange. Then, the toilet is set into place.

The closet bolts should be made of brass or some other material that will resist corrosive action. The closet bolts are tightened until the toi-

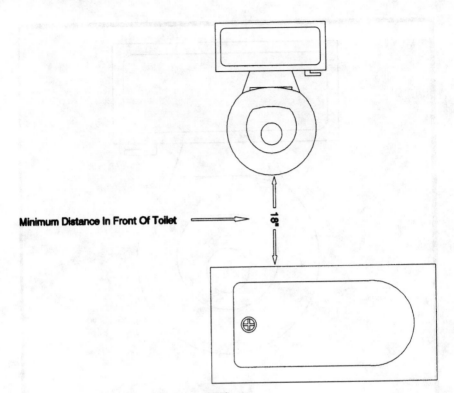

Minimum Distance In Front Of Toilet ⟶ 18"

Figure 7.9 Minimum distance in front of WC.

4981 Siphon Jet Extended Lip
• Wall hung urinal.

Figure 7.10 Urinal. (*Courtesy Universal-Rundle Corporation*)

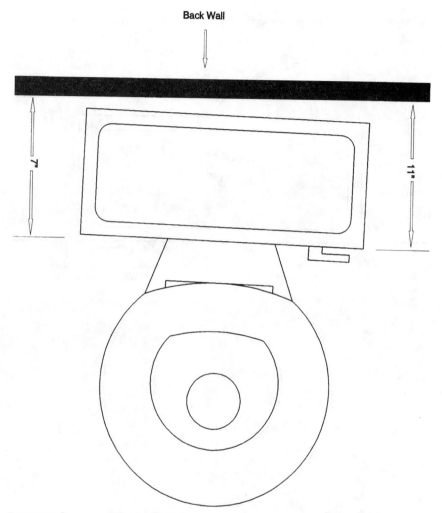

Back Wall

7"   11"

Figure 7.11 Improper toilet alignment.

let will not move from side to side or front to back. In some cases, a flange is not used, in which case the toilet should be secured with corrosion-resistant lag bolts.

When toilets or other fixtures are being mounted on a wall, the procedure is a little different. The fixture must be installed on, and supported by, an approved hanger. These hangers are normally packed with the fixture. The hanger must assume the weight placed in and on the fixture itself to avoid stress on the fixture.

In the case of a wall-hung toilet, the hanger usually has a pattern of bolts extending from the hanger to a point outside of the wall. The hanger is concealed in the wall cavity. A watertight joint is made at the

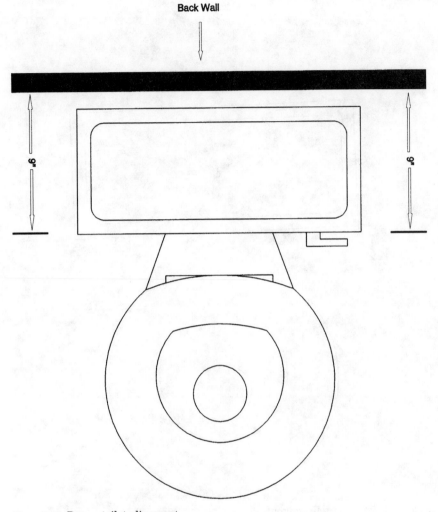

Figure 7.12 Proper toilet alignment.

point of connection, usually with a gasket ring, and the wall-hung toilet is bolted to the hanger.

With lavatories, the hanger is usually mounted on the outside surface of the finished wall. A piece of wood blocking is typically installed in the wall cavity to allow a solid surface for mounting the bracket. The bracket is normally secured to the blocking with lag bolts. The hanger is put in place and lag bolts are screwed through the bracket and finished wall into the wood blocking. Then, the lavatory is hung on the bracket.

The space where the lavatory meets the finished wall must be sealed. This is true of all fixtures coming into contact with walls, floor,

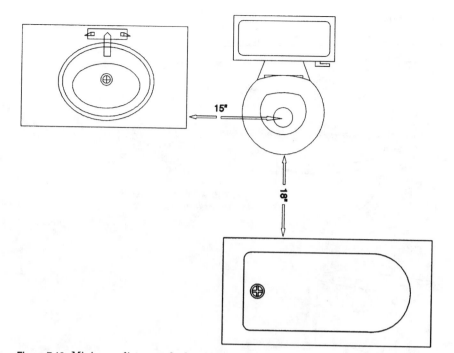

Figure 7.13 Minimum distances for legal layout.

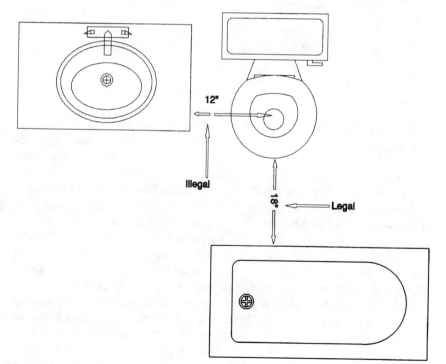

Figure 7.14 Illegal fixture spacing.

Figure 7.15 Wall-hung toilet. (*Courtesy Crane Plumbing*)

Figure 7.16 Floor-mount toilet. (*Courtesy Crane Plumbing*)

or cabinets. The crevice caused by the fixture meeting the finished surface must be sealed to protect against water damage. A caulking compound, such as silicone, is normally used for this purpose. This seal does more than prevent water damage. It eliminates hard-to-clean areas and makes the plumbing easier to keep free of dirt and germs.

When bathtubs are installed, they must be installed level, and they must be properly supported. The support for most one-piece units is the floor. These units are made to be set into place, leveled, and secured. Other types of tubs, like cast-iron tubs, require more support than the floor will give. They need a ledger or support blocks placed under the rim, where the edge of the tub meets the back wall.

The ledger can be a piece of wood, like a wall stud. The ledger should be about the same length as the tub. This ledger is installed horizontally and level. It should be at a height that will support the tub in a level fashion or with a slight incline, so excess water on the rim of the tub will run back into the tub. The ledger is nailed to wall studs.

If blocks are used, they are cut to a height that will put the bathtub into the proper position. Then, the blocks are placed at the two ends, and often in the middle, of where the tub will sit. The blocks should be installed vertically and nailed to the stud wall.

When the tub is set into place, the rim, at the back wall, rests on the blocks or ledger for additional support. This type of tub has feet on the bottom so that the floor supports most of the weight. The edges where the tub meets the walls must be caulked. If shower doors are installed on a bathtub or shower, they must meet safety requirements set forth in the building codes.

Showers today are usually one-piece units. These units are meant to sit in their place, be leveled, and be secured to the wall. The securing process for one-piece showers and bathtubs is normally accomplished by placing nails or screws through a nailing flange, which is molded as part of the unit, into the stud walls. If only a shower base is being installed, it must also be level and secure. Now, let's look at some of the many other regulations involved in installing plumbing fixtures.

The Facts about Fixture Installations

When it is time to install fixtures, there are many rules and regulations to adhere to. Water supply is one issue. Access is another. Air gaps and overflows are factors. There are a host of requirements governing the installation of plumbing fixtures. We will start with the fixtures most likely to be found in residential homes. Then, we will look at the fixtures normally associated with commercial applications.

Typical Residential Fixture Installation

Typical residential fixture installations could include everything from hose bibbs to bidets. This section is going to take each fixture that could be considered a typical residential fixture and tell you more about how they must be installed.

With most plumbing fixtures you have water coming into the fixture and water going out of the fixture. The incoming water lines must be protected against freezing and back-siphonage. Freeze protection is usually accomplished with the placement of the piping. In cold climates it is advisable to avoid putting pipes in outside walls. Insulation is often applied to water lines to reduce the risk of freezing. Back-siphonage is typically avoided with the use of air gaps and back-flow preventers.

Some fixtures, like lavatories and bathtubs, are equipped with overflow routes. These overflow paths must be designed and installed to prevent water from remaining in the overflow after the fixture is drained. They must also be installed in a manner that back-siphonage cannot occur. This normally means nothing more than having the faucet installed so that it is not submerged in water if the fixture floods. By keeping the faucet spout above the high-water mark, you have created an air gap. The path of a fixture's overflow must carry the overflowing water into the trap of the fixture. This should be done by integrating the overflow path with the same pipe that drains the fixture.

Bathtubs must be equipped with wastes and overflows. Zone one and zone three require these wastes and overflows to have a minimum diameter of 1½ in. The method for blocking the waste opening must be approved. Common methods for holding water in a tub include the following:

- Plunger-style stoppers
- Lift and turn stoppers
- Rubber stoppers
- Push and pull stoppers

Some fixtures, like hand-held showers pose special problems. Since the shower is on a long hose, it could be dropped into a bathtub full of water. If a vacuum was formed in the water pipe while the shower head was submerged, the unsanitary water from the bathtub could be pulled back into the potable water supply. This is avoided with the use of an approved back-flow preventer.

When a drainage connection is made with removable connections, like slip nuts and washers, the connection must be accessible. This normally isn't a problems for sinks and lavatories, but it can create some problems with bathtubs. Many builders and home buyers despise having an ugly access panel in the wall where their tub waste is located. To eliminate the need for this type of access, the tub waste can be connected with permanent joints. This could mean soldering a brass tub waste or gluing a plastic one. But, if the tub waste is connected with slip nuts, an access panel is required.

Washing machines generally receive their incoming water from boiler drains or laundry faucets. There is a high risk of a cross-connection when these devices are used with an automatic clothes washer. This type of connection must be protected against back-siphonage. The drainage from a washing machine must be handled by an indirect waste. An air break is required and is usually accomplished by placing the washer's discharge hose into a 2-in pipe as an indirect-waste receptor. The water supply to a bidet must also be protected against back-siphonage.

Dishwashers are another likely source of back-siphonage. These appliances must be equipped with either a back-flow protector or an air gap that is installed on the water-supply piping. The drainage from dishwashers is handled differently in each zone.

Zone one requires the use of an air gap on the drainage of a dishwasher. These air gaps are normally mounted on the countertop or in the rim of the kitchen sink. The air gap forces the waste discharge of the dishwasher through open air and down a separate discharge hose. This eliminates the possibility of back-siphonage or a back-up from the drainage system into the dishwasher.

Zone two requires dishwasher drainage to be separately trapped and vented or to be discharged indirectly into a properly trapped and vented fixture.

Zone three allows the discharge hose from a dishwasher to enter the drainage system in several ways. It may be individually trapped. It may discharge into a trapped fixture. The discharge hose could be connected to a wye tailpiece (Figs. 7.17 and 7.18) in the kitchen sink drainage. Further, it may be connected to the waste connection provided on many garbage disposers.

While we are on the subject of garbage disposers, be advised that garbage disposers require a drain of at least 1½ in and must be

Figure 7.17 Wye tailpiece.

Figure 7.18 Dishwasher drain adapter.

trapped. It may seem to go without saying, but garbage disposers must have a water source. This doesn't mean you have to pipe a water supply to the disposer; a kitchen faucet provides adequate water supply to satisfy the code.

Floor drains must have a minimum diameter of 2 in. Remember, piping run under a floor may never be smaller than 2 in in diameter. Floor drains must be trapped, usually must be vented, and must be equipped with removable strainers. It is necessary to install floor drains so that the removable strainer is readily accessible.

Laundry trays are required to have 1½-in drains. These drains should be equipped with cross-bars (Fig. 7.19) or a strainer. Laundry trays may act as indirect-waste receptors for clothes washers. In the case of a multiple-bowl laundry tray, the use of a continuous waste is acceptable.

Lavatories are required to have drains of at least 1¼ in in diameter. The drain must be equipped with some device to prevent foreign objects from entering the drain. These devices could include pop-up assemblies, cross-bars, or strainers.

When installing a shower, it is necessary to secure the pipe serving the shower head with water. This riser is normally secured with a drop-ear ell and screws. Figure 7.20 shows this type of installation. It is, however, acceptable to secure the pipe with a pipe clamp.

When we talk of showers here, we are speaking only of showers, not tub-shower combinations. The use of tub-shower combinations confuses many people. A shower has different requirements than those of a tub-shower combination. A shower drain must have a diameter of at least 2 in. The reason for this is simple. In a tub-shower combination, a 1½-in drain is sufficient, because the walls of the bathtub will retain

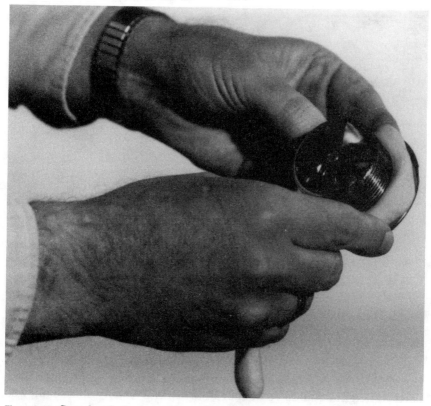

Figure 7.19 Cross-bar drain.

water until the smaller drain can remove it. A shower doesn't have high retaining walls; therefore, a larger drain is needed to clear the shower base of water more quickly. Shower drains must have removable strainers. The strainers should have a diameter of at least 3 in.

In zone three, all showers must contain a minimum of 900 in^2 of shower base. This area must not be less than 30 in in any direction. These measurements must be taken at the top of the threshold, and they must be interior measurements. A shower advertised as a 30-in shower may not meet code requirements. If the measurements are taken from the outside dimensions, the stall will not pass muster. There is one exception to the above ruling. Square showers with a rough-in of 32 in may be allowed. But the exterior of the base may not measure less than 31½ in.

Zone one requires the minimum interior area of a shower base to be at least 1024 in^2. When determining the size of the shower base, the measurements should be taken from a height equal to the top of the

Figure 7.20 Drop-ear ell.

threshold. The minimum size requirements must be maintained for a vertical height equal to 70 in above the drain. The only objects allowed to protrude into this space are grab bars, faucets, and shower heads.

The waterproof wall enclosure of a shower or a tub-shower combination must extend from the finished floor to a height of no less than 6 ft. Another criteria for these enclosures is that they must extend at least 70 in above the height of the drain opening. The enclosure walls must be at the higher of the two determining factors. An example of when this might come into play is a deck-mounted bathing unit. With a tub mounted in an elevated platform, an enclosure that extends 6 ft above the finished floor might not meet the criteria of being 70 in above the drain opening.

Although not as common as they once were, built-up shower stalls are still popular in high-end housing. These stalls typically use a concrete base, covered with tile. You may never install one of these classic shower bases, but you need to know how, just in case the need arises. These bases are often referred to as shower pans. Cement is poured into the pan to create a base for ceramic tile.

Before these pans can be formed, attention must be paid to the surface that will be under the pan. The subfloor, or other supporting surface, must be smooth and able to accommodate the weight of the shower. When the substructure is satisfactory, you are ready to make your shower pan.

Shower pans must be made from a waterproof material. In the old days, these pans were made of lead or copper. Today, they are generally made with coated papers or vinyl materials. These flexible materials make the job much easier. When forming a shower pan, the edges of the pan material must extend at least 2 in above the height of the threshold. Zone one requires the material to extend at least 3 in above the threshold. The pan material must also be securely attached to the stud walls.

Zone one goes deeper with its shower regulations. In zone one, the shower threshold must be 1 in lower than the other sides of the shower base, but the threshold must never be lower than 2 in. The threshold must also never be higher than 9 in. When installed for handicap facilities, the threshold may be eliminated.

Zone one goes on to require the shower base to slope toward the drain with a minimum pitch of ¼ in/ft, but not more than ½ in/ft. The opening into the shower must be large enough to accept a shower door with minimum dimensions of 22 in.

The drains for this type of shower base are new to many young plumbers; they may attempt to use standard shower drains for these types of bases, which you cannot do if you don't want the pan to leak. This type of shower base requires a drain that is similar to some floor drains.

The drain must be installed in a way that will not allow water that might collect in the pan to seep around the drain and down the exterior of the pipe. Any water entering the pan must go down the drain. The proper drain will have a flange that sits beneath the pan material, which will be cut to allow water into the drain. Then, another part of the drain is placed over the pan material and bolted to the bottom flange. The compression of the top piece and the bottom flange, with the pan material wedged between them, will create a watertight seal. Then, the strainer portion of the drain will screw into the bottom flange housing. Since the strainer is on a threaded extension, it can be screwed up or down to accommodate the level of the finished shower pan.

Sinks are required to have drains with a minimum diameter of 1½ in. Strainers or cross-bars are required in the sink drain. If you look, you will see that basket strainers have the basket part, as a strainer, and cross-bars below the basket. This provides protection from foreign objects even when the basket is removed. If a sink is equipped with a garbage disposer, the drain opening in the sink should have a diameter of at least 3½ in.

Toilets installed in zone three are required to be water-saver models. The older models that use 5 gallons (gal) per flush are no longer allowed in zone three for new installations.

The seat on a residential water closet must be smooth and sized for the type of water closet it is serving. This usually means that the seat will have a round front.

The fill valve or ballcock for toilets must be of the antisiphon variety. There are still older ballcocks being sold that are not of the antisiphon style. Just because these units are available doesn't make them acceptable. Don't use them; you will be putting your license and yourself on the line.

Toilets of the flush-tank type are required to be equipped with overflow tubes, which do double duty as refill conduits. The overflow tube must be large enough to accommodate the maximum water intake entering the water closet at any given time.

Whirlpool tubs must be installed as recommended by the manufacturer. All whirlpool tubs should be installed to allow access to the unit's pump. The pump's drain should be pitched to allow the pump to empty its volume of water when the whirlpool is drained. The whirlpool pump should be positioned above the fixture's trap.

All plumbing faucets and valves using both hot and cold water must be piped in a uniform manner. This manner calls for the hot water to be piped to the left side of the faucet or valve. Cold water should be piped to the right side of the faucet or valve. This uniformity reduces the risk of unwarranted burns from hot water.

In zone three, valves or faucets used for showers must be designed to provide protection from scalding. This means that any valve or faucet used in a shower must be pressure-balanced or contain a thermostatic-mixing valve. The temperature control must not allow the water temperature to exceed 110°F. This provides safety, especially to the elderly and the very young, against scalding injuries from the shower. Zones one and two do not require these temperature-controlled valves in residential dwellings. When zone one requires temperature-controlled shower valves, the maximum allowable temperature is 120°F.

Commercial Fixture Applications

Drinking fountains are a common fixture in commercial applications. Restaurants use garbages disposers that are so big it can take two plumbers to move them. Gang showers are not uncommon in school gyms and health clubs. Urinals are another common commercial fixture. Then, there are water closets. Water closets are in homes, but the ones installed for commercial applications often differ from residential toilets. Special fixtures and applications exist for some unusual plumbing fixtures, like baptismal pools in churches. This section is going to take you into the commercial field and show you how plumbing needs vary from residential uses to commercial applications.

Let's start with drinking fountains and water coolers. The main fact to remember about water coolers and fountains is this: they are not allowed in toilet facilities. You may not install a water fountain in a room that contains a water closet. If the building for which a plumbing diagram is being designed will serve water, such as a restaurant, or if the building will provide access to bottled water, drinking fountains and water coolers may not be required.

Commercial garbage disposers can be big. These monster grinding machines require a drain with a diameter of no less than 2 in. Commercial disposers must have their own drainage piping and trap. As with residential disposers, commercial disposers must have a cold-water source. In zone two, the water source must be of an automatic type. These large disposers may not be connected to a grease interceptor.

Garbage-can washers are not something you will find in the average home, but they are not uncommon in commercial applications. Due to the nature of this fixture, the water supply to the fixture must be protected against back-siphonage. This can be done with either a backflow preventer or an air gap. The waste pipe from these fixtures must have individual traps. The receptor that collects the residue from the garbage-can washer must be equipped with a removable strainer, capable of preventing the entrance of large particles into the sanitary drainage system.

Special fixtures are just that, special. Fixtures that might fall into this category include church baptismal pools, swimming pools, fish ponds, and other such arrangements. The water pipes to any of these special fixtures must be protected against back-siphonage.

Showers for commercial or public use can be very different from those found in a residence. It is not unusual for showers in commercial-grade plumbing to be gang showers. This amounts to one large shower enclosure with many shower heads and shower valves. In gang showers, the shower floor must be properly graded toward the shower drain or drains. The floor must be graded in a way to prevent water generated at one shower station from passing through the floor area of another shower station.

The methods employed to divert water from each shower station to a drain are up to the designer, but it is imperative that water used by one occupant may not pass into another bather's space. Zone one requires the gutters of gang showers to have rounded corners. These gutters must have a minimum slope toward the drains of 2 percent. The drains in the gutter must not be more than 8 ft from side walls and not more than 16 ft apart.

Urinals are not a common household item, but they are typical fixtures in public toilet facilities. The amount of water used by a urinal, in a single flush, should be limited to a maximum of 1½ gal. Water supplies to urinals must be protected from back-flow. Only one urinal may be flushed by a single flush valve. When urinals are used, they must not take the place of more than one-half of the water closets normally required. Public-use urinals are required to have a water trap seal that is visible and unobstructed by strainers.

Floor and wall conditions around urinals are another factor to be considered. These areas are required to be waterproof and smooth. They must be easy to clean, and they may not be made from an absorbent material. In zone three, these materials are required around a urinal in several directions. They must extend to at least 1 ft on each side of the urinal. This measurement is taken from the outside edge of the fixture. The material is required to extend from the finished floor to a point 4 ft off the finished floor. The floor under a urinal must be made of this same type of material, and the material must extend to a point at least 1 ft in front of the farthest portion of the urinal.

Commercial-grade water closets can present some of their own variations on residential requirements. The toilets used in public facilities must have elongated bowls. These bowls must be equipped with elongated seats. Further, the seats must be hinged and they must have open or split fronts.

Flush valves are used almost exclusively with commercial-grade fixtures (Fig. 7.21). They are used on water closets, urinals, and some special sinks. If a fixture depends on trap siphonage to empty itself, it must be equipped with a flush valve or a properly rated flush tank. These valves or tanks are required for each fixture in use.

4347 Mercury
- 18″ rim height.
- Floor mount bowl.
- 12″ rough-in.
- 3.5 G.P.F.

Figure 7.21 Flush-valve toilet.

Flush valves must be equipped with vacuum breakers that are accessible. Flush valves, in zone three, must be rated as water-conserving valves. These valves must be able to be regulated for water pressure, and they must open and close fully. If water pressure is not sufficient to operate a flush valve, other measures, such as a flush tank, must be incorporated into the design. All manually operated flush tanks should be controlled by an automatic filler, designed to refill the flush tank after each use. The automatic filler will be equipped to cut itself off when the trap seal is replenished and the flush tank is full. If a flush tank is designed to flush automatically, the filler device will be controlled by a timer.

Special Fixtures

There is an entire group of special fixtures that are normally found only in facilities providing health care. The requirements for these fixtures are extensive. While you may never have a need to work with these specialized fixtures, you should know the code requirements for them. This section is going to provide you with the information you may need.

Many special fixtures are required to be made of materials providing a higher standard than normal fixture materials. They may be required to endure excessive heat or cold. Many of these special fixtures are also required to be protected against back-flow. The fear of back-flow extends to the drainage system, as well as to the potable water supply. All special fixtures must be of an approved type.

Sterilizers

Any concealed piping that serves special fixtures and that may require maintenance or inspection must be accessible. All piping for sterilizers must be accessible. Steam piping to a sterilizer should be installed with a gravity system to control condensation and to prevent moisture from entering the sterilizer. Sterilizers must be equipped with a means to control the steam vapors. The drains from sterilizers are to be piped as indirect wastes. Sterilizers are required to have leak detectors. These leak detectors are designed to expose leaks and to carry unsterile water away from the sterilizer. The interior of sterilizers may not be cleaned with acid or other chemical solutions while the sterilizers are connected to the plumbing system.

Clinical sinks

Clinical sinks are sometimes called bedpan washers. Clinical sinks are required to have an integral trap. The trap seal must be visible and the contents of the sink must be removed by siphonic or blow-out action. The trap seal must be automatically replenished, and the sides of the fixture must be cleaned by a flush rim at every flushing of the sink. These special fixtures are required to connect to the drain waste and vent (DWV) system in the same manner as a water closet. When clinical sinks are installed in utility rooms, they are not meant to be a substitute for a service sink. On the other hand, service sinks may never be used to replace a clinical sink. Devices for making or storing ice shall not be placed in a soiled utility room.

Vacuum fluid-suction systems

Vacuum system receptacles are to be built into cabinets or cavities, but they must be visible and readily accessible. Bottle suction systems used for collecting blood and other human fluids must be equipped with overflow prevention devices at each vacuum receptacle. Secondary safety receptacles are recommended as an additional safeguard. Central fluid-suction systems must provide continuous service. If a central suction system requires periodic cleaning or maintenance, it must be installed so that it can continue to operate, even while cleaning or maintenance is being performed. When central systems are installed in hospitals, they must be connected to emergency power facilities. The vent discharge from these systems must be piped separately to the outside air, above the roof of the building.

Waste originating in a fluid suction system that is to be drained into the normal drainage piping must be piped into the drainage system with a direct-connect, trapped arrangement; an indirect-waste connection of this type of unit is not allowed.

Piping for these fluid suction systems must be noncorrosive and have a smooth interior surface. The main pipe shall have a diameter of no less than 1 in. Branch pipes must not be smaller than ½ in. All piping is required to have accessible cleanouts and must be sized according to manufacturer's recommendations. The air flow in a central fluid-suction system should not be allowed to exceed 5000 ft/min.

Special vents

Institutional plumbing uses different styles of vents for some equipment than what is encountered with normal plumbing. One such vent is called a local vent. One example of use for a local vent pertains to bedpan washers. A bedpan washer must be connected to at least one vent, with a minimum diameter of 2 in and that vent must extend to the outside air, above the roof of the building.

These local vents are used to vent odors and vapors. Local vents may not tie in with vents from the sanitary plumbing or sterilizer vents. In multistory buildings, a local vent stack may be used to collect the discharge from individual local vents for multiple bedpan washers, located above each other. A 2-in stack can accept up to three bedpan washers. A 3-in stack can handle six units, and a 4-in stack will accommodate up to twelve bedpan washers. These local vent stacks are meant to tie into the sanitary drainage system, and they must be vented and trapped if they serve more than one fixture.

Each local vent must receive water to maintain its trap seal. The water source shall come from the water supply for the bedpan washer being served by the local vent. A minimum of ¼-in tubing shall be run to the local vent, and it shall discharge water into the vent each time the bedpan washer is flushed.

Vents serving multiple sterilizers must be connected with inverted wye fittings, and all connections must be accessible. Sterilizer vents are intended to drain to an indirect waste. The minimum diameter of a vent for a bedpan sterilizer shall be 1½ in. When serving a utensil sterilizer, the minimum vent size shall be 2 in. Vents for pressure-type sterilizers must be at least 2½ in in diameter. When serving a pressure instrument sterilizer, a vent stack must be at least 2 in in diameter. Up to two sterilizers of this type may be on a 2-in vent. A 3-in stack can handle four units.

Water supply

Hospitals are required to have at least two water services. These two water services may, however, connect to a single water main. Hot water must be made available to all fixtures, as required by the fixture manufacturer. All water heaters and storage tanks must be of a type approved for the intended use.

Zone two requires the hot-water system to be capable of delivering 6½ gal of 125°F water per hour for each bed in a hospital. Zone two further requires hospital kitchens to have a hot water supply of 180°F water equal to 4 gal/h for each bed. Laundry rooms are required to have a supply of 180°F water at a rate of 4½ gal/h, for each bed. Zone two continues its hot-water regulations by requiring hot-water storage tanks to have capacities equal to no less than 8 percent of the water heating capacity.

Zone two continues with its hot-water requirements by dictating the use of copper in submerged steam heating coils. If a building is higher than three levels, the hot-water system must be equipped to circulate. Valves are required on the water distribution piping to fixture groups.

Back-flow prevention

When back-flow prevention devices are installed, they must be installed at least 6 in above the flood-level rim of the fixture. In the case of hand-held showers, the height of installation shall be 6 in above the highest point at which the hose can be used.

In most cases, hospital fixtures will be protected against back-flow by the use of vacuum breakers. However, a boiling-type sterilizer should be protected with an air gap. Vacuum suction systems may be protected with either an air gap or a vacuum breaker.

This has been a long chapter, but it was necessary to give you all the pertinent details on fixtures. As you now know, fixtures are not as simple as they may first appear. There are numerous regulations to learn and apply when installing plumbing fixtures. Your local jurisdiction may require additional or different code compliance. As always, check with your local authority before installing plumbing.

Reminder Notes

Zone one

1. Minimum fixture requirements vary from other zones.

2. Tub wastes and overflows must have a minimum diameter of 1½ in.

3. Dishwashers must drain through an air gap.

4. Requirements for shower stalls vary between the zones.

Zone two

1. Minimum fixture requirements vary from other zones.

2. Dishwashers must drain into a trap that is vented and used only for the dishwasher waste, or the waste hose must discharge indirectly into the piping of a properly trapped and vented fixture.

3. Commercial garbage disposers must be equipped with an automatic water source.

4. See the text for requirements on hot water in health care facilities.

Zone three

1. Minimum fixture requirements vary from other zones.

2. Urinals must have minimum clearances as follows: side-wall clearance of at least 15 in, center-to-center clearance of at least 30 in, and front clearance of at least 18 in.

3. Privacy stalls for water closets must have minimum dimensions of 30 in in width and 60 in in depth.

4. Tub wastes and overflows must have a minimum diameter of 1½ in.
5. Requirements for shower stalls vary between the zones.
6. Toilets are required to be water-saver models.
7. All shower valves must be of the type to prevent scalding.
8. Urinals may not use more than 1½ gal of water for a single flush.
9. Flush-valves must be of a water-conserving type.

8

Potable Water Systems

Potable water is essential to life. Water is often taken for granted in today's society, but it shouldn't be. Without good water, we would all die. Part of a plumber's responsibility is to provide safe drinking water. The requirements for making sure water is safe to drink are many. This chapter deals with the installation of potable water systems. It details approved types of piping and installation methods. These guidelines must be followed to ensure the health of our nation.

When we talk of potable water, we are speaking of water that is safe for drinking, cooking, and bathing, among other uses. Any building intended for human occupancy that has plumbing fixtures installed must have a potable water supply. If a building that has plumbing fixtures installed is intended for year-round habitation or is meant as a place of work for employees, both hot and cold potable water must be made available. All fixtures that are intended for use in bathing, drinking, cooking, and food processing or for creating medically related products must have potable water available, and only potable water. It is permissible to use nonpotable water for flushing toilets and urinals. Now that you know what must be equipped with potable water, let's see how to get the water to the fixtures.

The Main Water Pipe

The main water pipe delivering potable water to a building is called a water service. A water-service pipe must have a diameter of at least ¾ in. The pipe must be sized according to code requirements to provide adequate water volume and pressure to the fixtures. You will learn more about sizing a water service later in the chapter.

Ideally, a water-service pipe should be run from the primary water source to the building in a private trench, which means a trench not used for any purpose, except for the water service. However, it is normally allowable to place the water service in the same trench used by a sewer or building drain when specific installation requirements are followed. The water-service pipe must be separated from the drainage pipe. The bottom of the water-service pipe may not be closer than 12 in to the drainage pipe at any point.

A shelf must be made in the trench to support the water service. The shelf, as shown in Fig. 8.1, must be made solid and stable, at least 12 in above the drainage pipe. It is not acceptable to have a water service located in an area where pollution is probable. A water service should never run through, above, or under a waste disposal system, such as a septic field.

If a water service is installed in an area subject to flooding, the pipe must be protected against flooding. Water services must also be protected against freezing. The depth of the water service will depend on the climate of the location. Check with your local code officer to see how deep a water-service pipe must be buried in your area. Care must be taken when backfilling a water-service trench. The backfill material must be free of objects, like sharp rocks, that may damage the pipe.

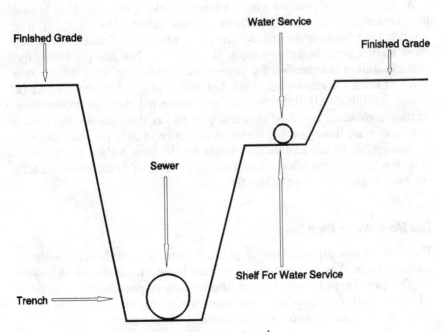

Figure 8.1 Water service and sewer in common trench.

When a water service enters a building through or under the foundation, the pipe must be protected by a sleeve. This sleeve is usually a pipe with a diameter at least twice that of the water service. Once through the foundation, the water service may need to be converted to an acceptable water-distribution pipe. As you learned in reading about approved materials, some materials approved for water-service piping are not approved for interior water distribution.

If the water-service pipe is not an acceptable water-distribution material, it must be converted to an approved material, generally within the first 5 ft of its entry into the building. Once inside a building, the maze of hot- and cold-water pipes are referred to as water-distribution pipes. Let's see what you need to know about water-distribution systems.

Water Distribution

Sizing a water-distribution system can become complicated. As we move through this chapter, we will leave sizing exercises until last. There are some rule-of-thumb methods that simplify the sizing of water-distribution pipes. Near the end of the chapter you will get information on sizing a system. For now, we will concentrate on other regulations.

Fixture supplies

Fixture supplies are the tubes or pipes that rise from the fixture branch, the pipe coming out of the wall or floor, to the fixture. In zone three, a fixture supply may not have a length of more than 30 in. The required minimum sizing for a fixture supply is determined by the type of fixture being supplied with water. Some common examples are listed in Table 8.1.

TABLE 8.1   Common Minimum Fixture-Supply Sizes

Fixture	Minimum supply size (in)
Bathtub	½
Bidet	⅜
Shower	½
Toilet	⅜
Lavatory	⅜
Kitchen sink	½
Dishwasher	½
Laundry tub	½
Hose bibb	½

Pressure-reducing valves

Pressure-reducing valves are required to be installed on water systems when the water pressure coming to the water-distribution pipes is in excess of 80 psi. The only time this regulation is generally waived is when the water service is bringing water to a device requiring high pressure.

Banging pipes

Banging pipes are normally the result of a water hammer. If you don't want complaining customers, avoid water hammers. You can avoid them in several ways. You might install air chambers above each faucet or valve (Fig. 8.2). Water-hammer arrestors are available and do a good job of controlling water hammer. Expansion tanks can also help with water hammer.

Water hammer is most prevalent around quick-closing valves, like ballcocks, and washing machine valves. Another way to reduce water hammer is to avoid long, straight runs of pipe (Fig. 8.3). By installing offsets in your water piping, you gradually break up the force of the water. By diminishing the force, you reduce water hammer.

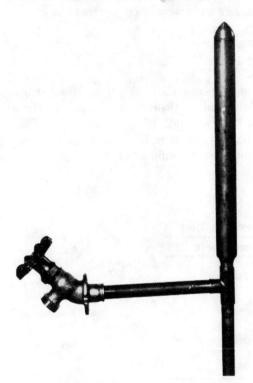

Figure 8.2 Air chamber.

Figure 8.3 Offset pipes.

Booster pumps

Not all water sources are capable of providing optimum water pressure. When this is the case, a booster pump may be required to increase water pressure. If water pressure fluctuates heavily, the water-distribution system must be designed to operate on the minimum water pressure anticipated.

When calculating the water pressure needs of a system, you can use information provided by your code book. There should be ratings for all common fixtures that will show the minimum pressure requirements for each type of fixture. A water-distribution system must be sized to operate satisfactorily under peak demands. Booster pumps are required to be equipped with low-water cut-offs. These safety devices are required to prevent the possibility of a vacuum, which may cause back-siphonage.

Water tanks

When booster pumps are not a desirable solution, water storage tanks are a possible alternative. Water storage tanks must be protected from contamination. They may not be located under soil or waste pipes. If the tank is of a gravity type, it must be equipped with overflow provisions.

The water supply to a gravity-style water tank must be automatically controlled. This may be accomplished with a ballcock or other suitable and approved device. The incoming water should enter the tank by way of an air gap that should be at least 4 in above the overflow.

Water tanks are also required to have provisions that allow them to be drained. The drain pipe must have a valve to prevent draining except when desired.

Pressurized water tanks

Pressurized water tanks are the type most commonly encountered in modern plumbing. These tanks are the type used with well systems. All pressurized water tanks should be equipped with a vacuum breaker. The vacuum breaker will be installed on top of the tank and should be at least ½ in in diameter. These vacuum breakers should be rated for proper operation up to maximum temperatures of 200°F.

It is also necessary to equip these tanks with pressure-relief valves. These safety valves must be installed on the supply pipe that feeds the tank or on the tank itself. The relief valve shall discharge when pressure builds to a point to endanger the integrity of the pressure tank. The valve's discharge must be carried by gravity to a safe and approved disposal location. The piping carrying the discharge of a relief valve may not be connected directly to the sanitary drainage system.

Supporting Your Pipe

The method used to support your pipes is regulated by the plumbing code. There are requirements for the types of materials you may use and how they may be used. Let's see what they are.

One concern with the type of hangers used is their compatibility with the pipe they are supporting. You must use a hanger that will not have a detrimental effect on your piping. For example, you may not use galvanized strap hanger to support copper pipe. As a rule of thumb, the hangers used to support a pipe should be made from the same material as the pipe being supported. For example, copper pipe should be hung with copper hangers (Figs. 8.4, 8.5, and 8.6). This eliminates the risk of a corrosive action between two different types of materials. If you are using a plastic hanger (Fig. 8.7) or plastic-coated one (Fig. 8.8), you may use it with all types of pipe.

The hangers used to support pipe must be capable of supporting the pipe at all times. The hanger must be attached to the pipe and to the member holding the hanger in a satisfactory manner. For example, it would not be acceptable to wrap a piece of wire around a pipe and then wrap the wire around the bridging between two floor joists. Hangers should be securely attached to the member supporting it. For example, a hanger should be attached to the pipe and then nailed to a floor joist.

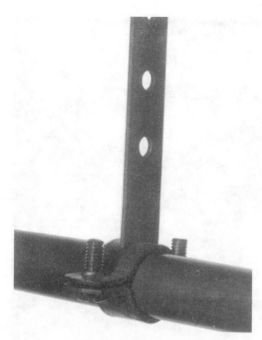

Figure 8.4 Copper pipe hanger.

Figure 8.5 Copper drive hanger.

Figure 8.6   Copper pipe clip.

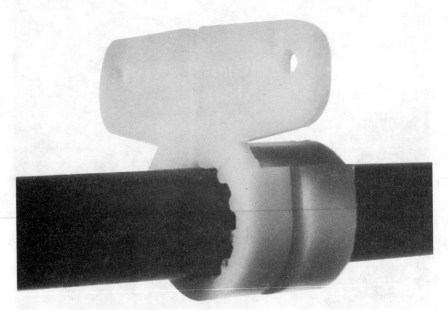

Figure 8.7   Plastic pipe hanger.

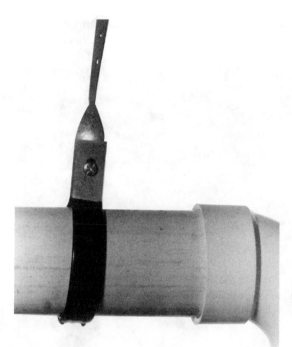

Figure 8.8 Plastic-coated pipe hanger.

The nails used to hold a hanger in place should be made of the same material as the hanger if corrosive action is a possibility.

Both horizontal and vertical pipes require support (Figs. 8.9 and 8.10). The intervals between supports will vary, depending upon the type of pipe being used and whether it is installed vertically or horizontally. The following examples will show you how often you must support the various types of pipes when they are hung horizontally; these examples are the maximum distances allowed between supports in zone three:

- ABS—every 4 ft
- Galvanized—every 12 ft
- Copper—every 6 ft
- CPVC—every 3 ft

- Cast iron—every 5 ft
- PVC—every 4 ft
- Brass pipe—every 10 ft
- PB pipe—every 32 in

When these same types of pipes are installed vertically in zone three, they must be supported at no less than the following intervals:

- ABS—every 4 ft
- Galvanized—every 15 ft
- Copper—every 10 ft
- Brass pipe—every 10 ft

- Cast iron—every 15 ft
- PVC—every 4 ft
- CPVC—every 3 ft
- PB pipe—every 4 ft

Figure 8.9 Vertical pipes supported.

Table 8.2 shows the support intervals for pipes installed in zone one. Table 8.3 shows the intervals allowed for pipes in zone two. Zone three's support requirements are shown in Table 8.4.

Water conservation

Water conservation continues to grow as a major concern. When setting the flow rates for various fixtures, water conservation is a factor. The flow rates of many fixtures must be limited to no more than 3 gpm. In zone three, these fixtures include the following:

- Showers
- Kitchen sinks
- Lavatories
- Other sinks

The rating of 3 gpm is based on a water pressure of 80 psi.

When installed in public facilities, lavatories must be equipped with faucets producing no more than ½ gpm. If the lavatory is equipped with a self-closing faucet, it may produce up to ¼ gpm per use. Water closets are restricted to a use of no more than 4 gal of water, and urinals must not exceed a usage of 1½ gal of water with each use.

Figure 8.10 Vertical pipes supported at a fixture.

Antiscald precautions

It is easy for the very young or the elderly to receive serious burns from plumbing fixtures. In an attempt to reduce accidental burns, it is required that mixed water to gang showers be controlled by thermo-static means or by pressure-balanced valves. All showers, except for showers in residential dwellings in zones one and two, must be

TABLE 8.2 Supporting Intervals for Supporting Water Pipe in Zone One

Type of pipe	Vertical support interval	Horizontal support interval (ft)
Threaded pipe (¾ in and smaller)	Every other story	10
Threaded pipe (1 in and larger)	Every other story	12
Copper tube (1½ in and smaller)	Every story, not to exceed 10 ft	6
Copper tube (2 in and larger)	Every story, not to exceed 10 ft	10
Plastic pipe	Not mentioned	4

TABLE 8.3   Support Intervals for Supporting Water Pipe in Zone Two

Type of pipe	Vertical support interval	Horizontal support interval (ft)
Threaded pipe	30 ft	12
Copper tube (1¼ in and smaller)	4 ft	6
Copper tube (1½ in)	Every story	6
Copper tube (larger than 1½ in)	Every story	10
Plastic pipe (2 in and larger)	Every story	4
Plastic pipe (1½ in and smaller)	4 ft	4

equipped with pressure-balanced valves or thermostatic controls. These temperature-control valves may not allow water with a temperature of more than 120°F to enter the bathing unit. In zone three, the maximum water temperature is 110°F. Zone three requires these safety valves on all showers.

Valve regulations

Gate valves and ball valves are examples of full-open valves, as required under valve regulations. These valves do not depend on rubber washers, and when they are opened to their maximum capacity, there is a full flow through the pipe. Many locations along the water-

TABLE 8.4   Pipe-Support Intervals in Zone Three

Type of vent pipe	Maximum distance of supports (ft)
Horizontal	
PB pipe	32 in
Lead pipe	Continuous
Cast iron	5 or at each joint
Galvanized	12
Copper tube (1¼ in)	6
Copper tube (1½ in and larger)	10
ABS	4
PVC	4
Brass	10
Aluminum	10
Vertical	
Lead pipe	4
Cast iron	15
Galvanized	15
Copper tubing	10
ABS	4
PVC	4
Brass	10
PB Pipe	4
Aluminum	15

distribution installation require the installation of full-open valves. Zone one requires these types of valves in the following locations:

- On the water service before and after the water meter
- On each water service for each building served
- On discharge pipes of water-supply tanks, near the tank
- On the supply pipe to every water heater, near the heater
- On the main supply pipe to each dwelling

In zone three, the locations for full-open valves are as follows:

- On the water-service pipe, near the source connections
- On the main water-distribution pipe, near the water service
- On water supplies to water heaters
- On water supplies to pressurized tanks, like well-system tanks
- On the building side of every water meter

Zone two requires full-open valves to be used in all water-distribution locations, except as cut-offs for individual fixtures, in the immediate area of the fixtures. There may be other local regulations that apply to specific building uses; check with your local code officer to confirm where full-open valves may be required in your system. All valves must be installed so that they are accessible.

Cut-offs

Cut-off valves do not have to be full-open valves. A stop-and-waste valve (Fig. 8.11) is an example of a cut-off valve that is not a full-open valve. Every sillcock must be equipped with an individual cut-off valve.

Figure 8.11 Stop-and-waste valve.

Appliances and mechanical equipment that have water supplies are required to have cut-off valves installed in the service piping. Generally, with only a few exceptions, cut-offs are required on all plumbing fixtures. Check with your local code officer for fixtures not requiring cut-off valves. All valves installed must be accessible.

Some Hard-Line Facts

- All devices used to treat or convey potable water must be protected against contamination.

- It is not acceptable to install stop-and-waste valves underground.

- If there are two water systems in a building, one potable, one nonpotable, the piping for each system must be marked clearly. The marking can be in the form of a suspended metal tag or a color-code. Zone two requires the pipe to be color coded and tagged. Nonpotable water piping should not be concealed.

- Hazardous materials, such as chemicals, may not be placed into a potable water system.

- Piping that has been used for a purpose other than conveying potable water may not be used as a potable water pipe.

- Water used for any purpose should not be returned to the potable water supply; this water should be transported to a drainage system.

Back-Flow Prevention

Back-flow and back-siphonage are both genuine health concerns. When a back-flow occurs, it can pollute entire water systems. Without back-flow and back-siphonage protection, municipal water services could become contaminated. There are many sources that are capable of deteriorating the quality of potable water.

Consider the example of someone using a water hose to spray insecticide on the grounds around a house. The device being used to distribute the insecticide is a bottle-type sprayer, attached to a typical garden hose. The bottle has just been filled for use with a poisonous bug killer. A telephone rings inside the home. The individual lays down the bottle sprayer and runs into the house to answer the phone.

While the individual is in the home and the bottle sprayer is connected, with the sillcock's valve open, a water main breaks. The back pressure caused by the break in the water main creates a vacuum. The vacuum sucks the poisonous contents of the bottle sprayer back into the potable water system. Now what? How far did the poison go? How much pipe and how many fixtures must be replaced before the water system can be considered safe? The lack of a back-flow preventer on the sillcock has created a nightmare. Human health and expensive financial considerations are at stake. A simple, inexpensive back-flow preventer could have avoided this potential disaster.

All potable water systems must be protected against back-siphonage and back-flow with approved devices. There are numerous types of devices available to provide this type of protection. The selection of devices will be governed by the local plumbing inspector. It is necessary to choose the proper device for the use intended.

An air gap is the most positive form of protection from back-flow. However, air gaps are not always feasible. Since air gaps cannot always be used, there are a number of devices available for the protection of potable water systems.

Some back-flow preventers are equipped with vents. When these devices are used, the vents must not be installed so that they may become submerged. It is also required that these units be capable of performing their function under continuous pressure.

Some back-flow preventers are designed to operate in a manner similar to an air gap. With these devices, when conditions occur that may cause a back-flow, the devices open and create an open air space between the two pipes connected to it. Reduced-pressure back-flow preventers perform this action very well. Another type of back-flow preventer that performs on a similar basis is an atmospheric-vent back-flow preventer.

Vacuum breakers are frequently installed on water heaters, hose bibbs, and sillcocks (Fig. 8.12). They are also generally installed on the faucet spout of laundry tubs. These devices either mount on a pipe or screw onto a hose connection. Some sillcocks are equipped with factory-installed vacuum breakers. These devices open, when necessary, and break any siphonic action with the introduction of air.

In some specialized cases a barometric loop is used to prevent back-siphonage. In zone three, these loops must extend at least 35 ft high and can only be used as vacuum breakers. These loops are effective because they rise higher than the point at which a vacuum suction can occur. Barometric loops work on the principle that by being 35 ft in height, suction will not be achieved.

Double-check valves are used in some instances to control back-flow. When used in this capacity, double-check valves must be equipped with approved vents. This type of protection would be used on a carbonated beverage dispenser, for example.

Back-flow preventers must be inspected from time to time. Therefore, they must be installed in accessible locations.

Some fixtures require an air gap as protection from back-flow. Some of these fixtures are lavatories, sinks, laundry tubs, bathtubs, and drinking fountains. This air gap is accomplished through the design and installation of the faucet or spout serving these fixtures.

Figure 8.12 Vacuum breaker.

When vacuum breakers are installed, they must be installed at least 6 in above the flood-level rim of the fixture. Vacuum breakers, because of the way they are designed to introduce air into the potable water piping, may not be installed where they may suck in toxic vapors or fumes. For example, it would not be acceptable to install a vacuum breaker under the exhaust hood of a kitchen range.

When potable water is connected to a boiler, for heating purposes, the potable water inlet should be equipped with a vented back-flow preventer. If the boiler contains chemicals in its water, the potable water connection should be made with an air gap or a reduced-pressure back-flow preventer.

Connections between a potable water supply and an automatic fire sprinkling system should be made with a check valve. If the potable water supply is being connected to a nonpotable water source, the connection should be made through a reduced-pressure back-flow preventer.

Lawn sprinklers and irrigation systems must be installed with back-flow prevention in mind. Vacuum breakers are a preferred method for back-flow prevention, but other types of back-flow preventers are allowed.

Hot-Water Installations

When hot-water pipe is installed, it is often expected to maintain the temperature of its hot water for a distance of up to 100 ft from the fixture it serves. If the distance between the hot-water source and the fixture being served is more than 100 ft, a recirculating system is frequently required. When a recirculating system is not appropriate, other means may be used to maintain water temperature. These means could include insulation or heating tapes. Check with your local code officer for approved alternates to a recirculating system, if necessary.

If a circulator pump is used on a recirculating line, the pump must be equipped with a cut-off switch. The switch may operate manually or automatically.

Water Heaters

The standard working pressure for a water heater is 125 psi. The maximum working pressure of a water heater is required to be permanently marked in an accessible location. Every water heater is required to have a drain, located at the lowest possible point on the water heater. Some exceptions may be allowed for very small, under-the-counter water heaters.

All water heaters are required to be insulated. The insulation factors are determined by the heat loss of the tank in an hour's time. These specifications are required of a water heater before it is approved for installation.

Relief valves are mandatory equipment on water heaters (Fig. 8.13). These safety valves are designed to protect against excessive temperature and pressure. The most common type of safety valve used will protect against both temperature and pressure, from a single valve. The blow-off rating for these valves must not exceed 210°F and 150 psi. As far as the rating for the pressure relief valve goes, the valve must not have a blow-off rating of more than the maximum working pressure of the water heater it serves; this is usually 125 psi.

When temperature and pressure relief valves are installed, their sensors should monitor the top 6 in of water in the water heater. There may not be any valves located between the water heater and the temperature and pressure relief valves.

The blow-off from relief valves must be piped down, to protect bystanders, in the event of a blow-off. The pipe used for this purpose must be rigid and capable of sustaining temperatures of up to 210°F. The discharge pipe must be the same size as the relief valve's discharge opening, and it must run, undiminished in size, to within 6 in of the

Figure 8.13 Temperature and pressure relief valve.

floor. If a relief valve discharge pipe is piped into the sanitary drainage system, the connection must be through an indirect waste. The end of a discharge pipe may not be threaded, and no valves may be installed in the discharge pipe.

When the discharge from a relief valve may damage property or people, safety pans should be installed. These pans typically have a minimum depth of 1½ in. Plastic pans are commonly used for electric water heaters, and metal pans are used for fuel-burning heaters. These pans must be large enough to accommodate the discharge flow from the relief valve.

The pan's drain may be piped to the outside of the building or to an indirect waste, where people and property will not be affected. The discharge location should be chosen so that it will be obvious to building occupants when a relief valve discharges. Traps should not be installed on the discharge piping from safety pans.

Water heaters must be equipped with an adjustable temperature control. This control is required to be automatically adjustable from the lowest to the highest temperatures allowed. Some locations restrict the maximum water temperature in residences to 120°F. There must be a switch supplied to shut off the power to electric water heaters. When the water heater uses a fuel, like gas, there must be a valve available to cut off the fuel source. Both the electric and fuel shut-offs must be able to be used without affecting the remainder of the building's power or fuel. All water heaters requiring venting must be vented in compliance with local code requirements.

Purging the System of Contaminants

When a potable water system has been installed, added to, or repaired, it must be disinfected. For years, this amounted to little more than running water through the system until it appeared clean. This is no longer enough. Under today's requirements, the system must undergo a true cleansing procedure.

The precise requirements for clearing a system of contaminants will be prescribed by the local health department or code enforcement office. Typically, it will require flushing the system with potable water until the water appears clean. Then, this action will be followed by a cleaning with a chlorine solution. The exact requirements for the mixture of chlorine and water will be provided by a local agency. The chlorine mixture will be introduced into the system and normally will be required to remain in the system for between 3 and 24 h.

After the chlorine has been in the system for the required time, the system will be flushed with potable water until there is no trace of chlorine remaining. Again, check with your local authorities on the exact specifications for purifying the potable water system.

Working with Wells

When you will be working with wells, or other private water sources, there are some rules you must follow. This section will tell you what you need to know when working with private water systems.

If a building does not have access to a public water source, it must depend on water from a private source. Typically, this source is a well. But, under some conditions, it could be a cistern, spring, or stream. If surface water is used as a potable water source, it must be tested and approved for use. For that matter, wells are generally required to be tested and approved.

The quality of water from a private source must meet minimum standards as potable water. This is determined through water tests. The determination of what potable water is is normally done by the local health department or some other local authority.

The quantity of water delivered from a well must also meet certain requirements. The well, or water source, must be capable of supplying enough water for the intended use of the system. For example, zone three rates a single-family home as requiring 75 gal of water a day for each occupant. Hospitals, on the other hand, are rated to require a minimum of between 150 and 250 gal of water each day for each bed in the hospital.

All private water sources must be protected from contamination. They must also be disinfected before being used. The protection from contamination can include several factors. For example, wells must have watertight caps. They may not be installed in an area where contamination is likely, such as being installed near a septic system. The height of a well casing should extend above the ground. All wells should be located above and upstream from any possible contaminating sources, such as a septic field.

Any well, whether drilled, dug, or driven, generally must not deliver water for potable use from a depth of less than 10 ft. There are rules governing the allowable distances between a private water source and possible pollution sources. The following examples show how far a private water source in zone three must be from a few of the possible polluting sources:

- Septic tank—25 ft
- Pasture land—100 ft
- Sewer—10 ft
- Barnyard—100 ft
- Underground disposal fields—50 ft

Construction Requirements for Wells

Wells must be installed to meet minimum standards. What follows is a description of the minimum requirement for the installation of various types of wells in zone three.

Dug and bored wells

Dug and bored wells are usually relatively shallow. Their casings are required to be made from waterproof concrete, corrugated metal pipe, galvanized pipe, or tile. These casings must extend to a minimum depth of 10 ft below the ground. The casing is required to extend below the water table. For example, if the well is 16 ft deep and the water table is 13 ft deep, the casing must extend at least 13 ft into the ground.

When wells are dug or bored, there is a large space between the well casing and the undisturbed earth. This space is required to be filled with a grout material. The grout must encompass the well casing and have a minimum thickness of 6 in. This helps to prevent surface water from entering the well. It is also necessary for the well casing to rise at least 6 in above the well platform.

Then, there is the cover. This type of well is usually large in diameter. The top of the well must be sealed with a watertight cover. Covers must overlap the sides of the well casing and extend downward for a minimum of 2 in. Concrete covers are common on this type of well.

Common practice with bored and dug wells is to have the water pipe, going from the well to the house, exit through the side of the casing. This is generally done below ground, below the regional frost line. Where this penetration of the well casing occurs, the hole must be sealed to prevent outside water from entering the well.

If a well is installed in an area subject to flooding, the well casing and cover must be designed and installed to withstand the forces associated with a flood. Grading of the ground surrounding a private water source may be required to divert run-off water from entering the potable water supply.

Drilled and driven wells

Drilled and driven wells are different from dug and bored wells. The diameters of these wells are much smaller, and drilled wells are often very deep. The casings for these wells must be made of steel or some other approved material, but they are usually made of steel (Fig. 8.14). The casing must extend at least 6 in above the well platform.

Figure 8.14 Steel well casing for a drilled well.

Grouting is required around the exterior of these casings. The grout material is required to a minimum depth of 10 ft or solid contact with rock, whichever comes first. The casing should extend into rock or well beyond the water table level.

Getting the water pipe from the well to a building, in a drilled well, is usually done through the side of the casing. Normal procedure calls for the use of a pitless adapter in these installations. Pitless adapters mount into the well casing and form a watertight seal. In any event, the casing must be sealed at any openings that might allow nonpotable water to enter the well.

The cover for a drilled well must be waterproof and will usually allow provisions for electrical wires connected to the pump. These wires must get from the submerged pump to the control box. In all cases, the cover must be designed and installed to prevent the influx of surface water into the well.

Well Pumps

The pumps used with potable water systems must meet minimum standards. This section identifies and explains these standards. Pumps must be approved for use. They must be readily accessible for service, maintenance, or repair. In flood-prone areas, pumps must be designed and installed to resist the potential detrimental effects of a flood. Water pumps are required to be capable of continuous operation.

In zone three, if a pump is installed in a home, it must be installed on an appropriate base. Some pumps are installed on brackets, connected to pressure supply tanks. If a pump is installed in a basement, it must be installed at least 18 in above the basement floor. It is not acceptable to install the pump in a pit or closer than 18 in to the finished floor level of a basement floor. This provision is meant to protect the pump from submersion, through basement flooding.

Pump Houses

When shallow-well pumps or two-pipe jet pumps are used, the pumps are sometimes placed in pump houses. These pump houses must be of approved construction. A building providing shelter to a water pump must be equipped to prevent the pump or related piping from freezing. Such an enclosure must also be provided with adequate drainage facilities to prevent water from rising over the pump and piping.

Sizing Potable Water Systems

About the only aspect of the potable water system that we have not covered is sizing. This section of the chapter is going to show you how to size potable water piping, but, be advised, this procedure is not always simple and requires concentration. If your mind is not fresh, leave this section for a later reading. If you are ready to learn how to size water pipe, get a pen and paper and get ready for one of the more complicated aspects of this book.

Some facets of potable water pipe sizing are not very difficult. Many times your code book will provide charts and tables to help you. Some of these graphics will detail precisely what size pipe or tubing is required. But, unfortunately, code books cannot provide concrete answers for all piping installations.

Many factors affect the sizing of potable water piping. The type of pipe used will have an influence on your findings. Some pipe materials have smaller inside diameters than others. Some pipe materials have a rougher surface or more restrictive fittings than others. Both of these factors will affect the sizing of a water system.

When sizing a potable water system, you must be concerned with the speed of the flowing water, the quantity of water needed, and the restrictive qualities of the pipe being used to convey the water. Most materials approved for potable water piping will provide a flow velocity of 5 ft per second (ft/s). The exception is galvanized pipe. Galvanized pipe will provide a speed of 8 ft/s.

It may be surprising that galvanized pipe allows a faster flow rate. This occurs because of the wall strength of galvanized pipe. In softer pipes, like copper, fast-moving water can essentially wear a hole in the pipe. These flow ratings are not carved in stone. You will find people who will argue for either a higher or a lower rating, but these ratings are in use with current plumbing codes.

When using the three factors previously discussed to determine pipe size, you must use math that you may not have seen since your school days, and you may not have seen it then. Here is an example of how a typical formula might look:

$$Y = XZ$$

where X = the water's rate of flow, in most cases, 5 ft/s
      Y = the quantity of water in the pipe
      Z = the inside diameter of the pipe

Since most plumbers will refuse to do this type of math, most code books offer alternatives. The alternatives are often in the form of tables or charts that show pertinent information on the requirements for pipe sizing.

The tables or charts you might find in a code book are likely to discuss the following: a pipe's outside diameter, a pipe's inside diameter, a flow rate for the pipe, and a pressure loss in the pipe over a distance of

100 ft. These charts or tables will be dedicated to a particular type of pipe. For example, there would be one table for copper pipe and another for PB pipe.

The information supplied in a ratings table for PB pipe might look like this:

- Pipe size is ¾ in.
- Inside pipe diameter is 0.715 in.
- Flow rate, at 5 ft/s, is 6.26 gpm.
- Pressure lost in 100 ft of pipe is 14.98.

This type of pipe sizing is most often done by engineers, not plumbers. When sizing a potable water system, the sizing exercise starts at the last fixture and works its way back to the water service.

Commercial jobs, where pipe sizing can get quite complicated, are generally sized by design experts. All a working plumber is required to do is install the proper pipe sizes in the proper locations and manner. For residential plumbing, where engineers are less likely to have a hand in the design, there is a rule-of-thumb method to sizing most jobs. In the average home, a ¾-in pipe is sufficient for the main artery of the water-distribution system. Normally, not more than two fixtures can be served by a ½-in pipe. With this in mind, sizing becomes simple.

A ¾-in pipe is normally run to the water heater, and it is typically used as a main water-distribution pipe. When nearing the end of a run, the ¾-in pipe is reduced to ½-in pipe, once there are no more than two fixtures to connect to. Most water services will have a ¾-in diameter, with those serving homes with numerous fixtures being a 1-in pipe. This rule-of-thumb sizing will work on almost any single-family residence.

The water supplies to fixtures are required to meet minimum standards. These sizes are derived from local code requirements. You simply find the fixture you are sizing the supply for and check the column heading for the proper size.

Most code requirements seem to agree that there is no definitive way to set a boiler-plate formula for establishing potable water pipe sizing. Code officers can require pipe sizing to be performed by a licensed engineer. In most major plumbing systems the pipe sizing is done by a design professional.

Code books give examples of how a system might be sized. But the examples are not meant as a code requirement. The code requires a water system to be sized properly. However, due to the complexity of the process, the books do not set firm statistics for the process. Instead, code books give parts of the puzzle, in the form of some minimum standards, but it is up to a professional designer to come up with an approved system.

Where does this leave you? Well, the sizing of a potable water system is one of the most complicated aspects of plumbing. Very few single-family homes are equipped with potable water systems designed by engineers. You now have a basic rule-of-thumb method for sizing small systems. Next, you will learn how to use the fixture-unit method of sizing.

The fixture-unit method is not very difficult, and it is generally acceptable to code officers. While this method may not be perfect, it is much faster and easier to use than the velocity method. Other than the additional expense in materials, you can't go wrong by oversizing your pipe. If in doubt on sizing, go to the next larger size. Now, let's see how you might size a single-family residence's potable water system using the fixture-unit method.

Most codes assign a fixture-unit value to common plumbing fixtures. To size with the fixture-unit method, you must establish the number of fixture units to be carried by the pipe. You must also know the working pressure of the water system. Most codes will provide guidelines for these two pieces of information.

For our example, we have a house with the following fixtures: three toilets, three lavatories, one bathtub-shower combination, one shower, one dishwasher, one kitchen sink, one laundry hook-up, and two sill-cocks. The water pressure serving this house is 50 psi. There is a 1-in water meter serving the house, and the water service is 60 ft in length. With this information and the guidelines provided by your local code, you can do a pretty fair job of sizing your potable water system.

The first step is to establish the total number of fixture units on the system. The code regulations will provide this information. In this case, the fixture-unit ratings are listed in Table 8.5. Use the table to find the total number of fixture units in the house. You have three toilets, that's 9 fixture units. The three lavatories add 3 fixture units. The tub-shower combination counts as 2 fixture units; the shower head over the bathtub doesn't count as an additional fixture. The shower has 2 fixture units. The dishwasher adds 2 fixture units and so does the kitchen sink. The laundry hook-up counts as 2 fixture units. Each sillcock is worth 3 fixture units. This house has a total fixture-unit load of 28.

Now you have the first piece of your sizing puzzle solved. The next step is to determine what size pipe to use for your number of fixture units. Refer to Table 8.6, it provides guidelines, similar to those provided in the plumbing codes, that pertain to pressure rating, pipe length, and fixture loads.

Our subject house has a water pressure of 50 psi. This pressure rating falls into the category allowed in Table 8.6. First, find the proper

TABLE 8.5   Equivalent Fixture-Unit Ratings for Sizing Example

Fixture	Fixture-unit rating combining hot and cold water
Toilet	3
Lavatory	1
Tub/shower combination	2
Shower	2
Dishwasher	2
Kitchen sink	2
Laundry hookup	2
Sillcock	3

TABLE 8.6   Sample Pressure and Pipe Chart for Sizing Exercise—Sizing Table for Water Pressure Ranging from 46 to 60 psi

Size of water meter and street service (in)	Size of water service and distribution pipes (in)	Fixture units and maximum length of water pipe					
		40 ft	60 ft	80 ft	100 ft	150 ft	200 ft
¾	½	9	8	7	6	5	4
¾	¾	27	23	19	17	14	11
¾	1	44	40	36	33	28	23
1	1	60	47	41	36	30	25
1	1¼	102	87	76	67	52	44

water meter size; the one you are looking for is 1 in. You will notice that a 1-in meter and a 1-in water service are capable of handling 60 fixture units, when the pipe is only running 40 ft. However, when the pipe length is stretched to 80 ft, the fixture load is dropped to 41. At 200 ft, the fixture rating is 25. What is it at 100 ft? At 100 feet, the allowable fixture load is 36. See, this type of sizing is not so hard.

Now, what does this tell us? Well, we know the water service is 60 ft long. Once inside the house, how far is it to the most remote fixture? In this case, the farthest fixture is 40 ft from the water-service entrance. This gives us a developed length of 100 ft, 60 ft for the water service and 40 ft for the interior pipe. Going back to Table 8.6, we see that for 100 ft of pipe, under the conditions in this example, we are allowed 36 fixture units. The house only has 28 fixture units, so, our pipe sizing is fine.

What would happen if the water meter was a ¾-in meter instead of a 1-in meter? With a ¾-in meter and a 1-in water service and main distribution pipe, we could have 33 fixture units. This would still be a suitable arrangement since we only have 28 fixture units. Could we use a ¾-in water-service and water-distribution pipe with the ¾-in meter? No, we couldn't. With all sizes set at ¾ in, the maximum number of fixture units allowed is 17.

In this example, the piping is oversized. But, if you want to be safe, use this type of procedure. If you are required to provide a riser diagram showing the minimum pipe sizing allowed, you will have to do a little more work. Once inside a building, water-distribution pipes normally extend for some distance, supplying many fixtures with water. As the distribution pipe continues on its journey, it reduces the fixture load as it goes.

For example, assume that the distribution pipe serves a full bathroom group within 10 ft of the water service. Once this group is served with water, the fixture-unit load on the remainder of water-distribution piping is reduced by 6 fixture units. As the pipe serves other fixtures, the fixture-unit load continues to decrease. So, it is feasible for the water-distribution pipe to become smaller as it goes along.

Table 8.6 can also be used to find a minimum pipe size. Let's take our same sample house and see how we could use smaller pipe. Okay, we know we need a 1-in water service. Once inside the foundation, the water service becomes the water-distribution pipe. The water heater is located 5 ft from the cold-water distribution pipe. The 1-in pipe will extend over the water heater and supply it with cold water. Then, there will be a hot-water distribution pipe originating at the water heater. Now you have two water-distribution pipes to size.

When sizing the hot- and cold-water pipes, you could make adjustments for fixture-unit values on some fixtures. For example, a bathtub is rated as 2 fixture units. Since the bathtub rating is inclusive of both hot and cold water, obviously the demand for just the cold-water pipe is less than that shown in our table. For simplicity's sake, the fixture units will not be broken down into fractions or reduced amounts. The example will assume that a bathtub requires 2 fixture units of hot water and 2 fixture units of cold water. However, you could reduce the amounts listed in the table by about 25 percent to obtain the rating for each hot- and cold-water pipe. For example, the bathtub, when being sized for only cold water, could take on a value of 1½ fixture units.

Now then, let's get on with the exercise. We are at the water heater. We ran a 1-in cold-water pipe overhead and dropped a ¾-in pipe into the water heater. What size pipe do we bring up for the hot water? First, count the number of fixtures that use hot water and assign them fixture-unit values. The fixtures using hot water are all fixtures except the toilets and sillcocks. The total count for hot-water fixture units is lucky number 13. From the water heater, our most remote hot-water fixture is 33 ft away.

What size pipe should we bring up from the water heater? By looking at the Table 8.6, find a distance and fixture-unit count that will work in this case. You would look under the 40-ft column, since our distance is less than 40 ft. When you look in the column, the first fixture-unit number you see is 9; this won't work. The next number is 27; this one will work because it is greater than the 13 fixture units we need. Looking across the table, you will see that the minimum pipe size to start with is a ¾-in pipe. Isn't it convenient that the water heater just happens to be sized for ¾-in pipe?

Okay, now we start our hot-water run with ¾-in pipe. As our hot-water pipe goes along the 33-ft stretch, it provides water to various fixtures. When the total fixture count remaining to be served drops to less than 9 fixture units, we can reduce the pipe to ½-in pipe. We can also run our fixture branches off of the main in ½-in pipe. We can do this because the highest fixture-unit rating on any of our hot-water fixtures is 2 fixture units. Even with a pipe run of 200 ft we can use ½-in pipe for up to 4 fixture units. Is this sizing starting to ring a bell? Remember the rule-of-thumb sizing given earlier? These sizing examples are making the rule-of-thumb method ring true.

With the hot-water sizing done, let's look at the remainder of the cold water. We have less than 40 ft to our farthest cold-water fixture. We branch off near the water-heater drop for a sillcock, and there is a full bathroom group within 7 ft of our water-heater drop. The sillcock branch can be ½-in pipe. The pipe going under the full bathroom group could probably be reduced to ¾-in pipe, but it would be best to run it as a 1-in pipe. However, after serving the bathroom group and the sillcock, how may fixture units are left. There are only 19 fixture units left. We can now reduce to ¾-in pipe, and when the demand drops to below 9 fixture units, we can reduce to ½-in pipe. All of our fixture branches can be run with ½-in pipe.

This is one way to size a potable water system that works, without driving you crazy. There may be some argument to the sizes given in these examples. The argument would be that some of the pipe is oversized, but when in doubt, go bigger. In reality, the cold-water pipe in the last example could probably have been reduced to ¾-in pipe where the transition was made from water-service to water-distribution pipe. It could have almost certainly been reduced to ¾-in pipe after the water-heater drop. Local codes will have their own interpretations of pipe sizing, but this method will normally serve you well.

9

After the Installation

After installing various phases of plumbing, the work must be inspected and approved. The inspections are generally performed by local plumbing inspectors. The permit holder is responsible for all costs and efforts required to test the plumbing. It is not uncommon to have as many as three or four inspections. These inspections might include inspection of water services, sewers, underground plumbing, rough-in plumbing, and final plumbing.

Before a plumber's job is finished, the work must be inspected and approved. Time and money spent on reinspections is lost time and lost profits. When you are installing plumbing for profit, it is especially important to get it right the first time. This chapter is going to take you through the general requirements for testing each plumbing phase.

It is often acceptable to test plumbing phases in sections. But, common procedure calls for testing entire phases simultaneously. Generally, air or water may be used for testing plumbing. Occasionally, special tests, like smoke or peppermint tests, will be required. Let's take a look at permissible ways to test your plumbing.

Testing Sewers

There are two common methods for testing building sewers. The first method uses water; the second uses air. In either case, the building sewer should be capped or plugged at the point where it will connect with the main sewer. Test-tee fittings are commonly installed in this portion of the sewer to allow for the test. Sewers must be tested, inspected, and approved before they are covered. Sewers should be covered by a minimum of 12 in of earth.

When testing with water, the sewer must be filled with water to a point equal to a 10-ft head. In simple terms, this means extending a pipe, like a cleanout riser, to a point 10 ft higher than the sewer. The pipe rising to allow for the 10-ft head should have water resting at its upper limit. The water must be visible.

When testing with water, water pressure must be maintained for at least 15 min before an official inspection is made. If the water level goes down, you've got a problem. All joints must be watertight.

If leaks are present, plan on cutting them out and replacing them. You should not patch the leaks with wax or glue. This is done from time to time, but it is not right. Also, don't try to pull a fast one on the inspector. Some plumbers put a plastic test cap on the pipe rising above the sewer, to give a false impression. With the plastic cap in place, the water level in the head riser would not fall, but the sewer had no water in it. The plastic cap, at the fitting entering the sewer is not visible, but a smart inspector will require you to release the test water so the flow of water can be seen and identified as having filled the sewer and test riser.

When testing with air, you must rig the sewer to accept a pressure gauge. The sewer must be pumped with air until the contents reach a pressure of at least 5 psi. If a mercury gauge is used, the pressure must balance 10 in of mercury. The time requirements for an air test are the same as those for a water test.

Testing the Water Service

The test of a water service is sometimes waived. If the water service comprises a single pipe, with no joints, a pressure test may not be required. If a test is required, the pipe can be tested with potable water or air. The water service must be tested, inspected, and approved before being buried. Water services must be located deep enough to prevent freezing.

Zones one and three require water services to be tested at a pressure equal to their maximum working pressure. Zone two requires the test pressure to be set at a pressure of at least 25 psi higher than the maximum working pressure.

Testing Groundworks

Underground plumbing is tested in essentially the same way that a building sewer is tested. An air pressure of 5 psi or a 10-ft head of water is required. When a mercury gauge is used, the test must balance a 10-in column of mercury. The test must be maintained for at least 15 min prior to inspection.

Testing the Drain Waste and Vent (DWV) Rough-In

All DWV rough-ins must be inspected before being concealed. When testing with air, vent terminals, fixture outlets, and the building drain must be capped or plugged (Fig. 9.1). The DWV system must be subjected to a 15-min test with either air or water. If air is used, the system must be tested with a minimum pressure of 5 psi or a 10-in column of mercury.

When testing a DWV system with water, the test-water level is usually required to extend to the top of the roof vents. Some areas will allow the test to terminate at the flood-level rim of the highest bathing unit in the premises, but normally, the water must be to the top of the vents.

During the DWV test, inspectors will look for pipe protection. When a pipe is installed in a way that it may be penetrated by nails or screws, the pipe must be protected with nail plates. If structural members have been substantially weakened by your plumbing installation, your job will not pass inspection. Pipe hangers will also be inspected.

Figure 9.1 Plastic test cap.

Testing the Water-Distribution Rough-In

All water-distribution pipes must be tested, inspected, and approved before being concealed (Fig. 9.2). The test pressure required for a potable water system is usually the same as the maximum working pressure for the system. However, zone two requires the test pressure to be 25 psi higher than the working pressure.

Another consideration in a water-pipe inspection includes pipe protection from punctures and freezing. Pipe hangers are also inspected. Back-flow preventers, air gaps, and all other code requirements are examined in these inspections, as they are with other inspections.

Testing the Final Plumbing

The job is not done until the final approval is issued from the code enforcement office. What is involved in a final plumbing inspection?

Figure 9.2 Test rig on water pipes.

Well, typically, the inspection is a matter of a visual tour of the plumbing. This tour normally includes the inspector's use and observation of all plumbing fixtures. For example, an inspector will test to see that the hot water is piped to the left side of a faucet. The inspector will check traps and other connections for leaks.

In the final inspection, inspectors put all the plumbing fixtures through their paces. Cut-offs are inspected, aerators are checked, backflow preventers are checked, fixtures are filled and drained, and water heaters may be tested. In general, all plumbing is checked to assure proper installation procedures and working conditions.

If an inspector has reason to suspect a plumbing system is not up to snuff, the inspector may require a smoke or peppermint test. These tests are designed to expose leaks in the DWV system. In these tests, all traps are filled with water. The DWV system is filled with a colored smoke or an oil of peppermint. When the smoke is visible at a vent or the peppermint is noticeable, the vents are capped. Then, the inspector will check each trap for evidence of a leak. The colored smoke or aromatic peppermint makes it easy to find traps that are not doing their jobs.

Interior Rain Leaders and Downspouts

Interior rain leaders and downspouts should be tested, inspected, and approved in the same manner used for DWV systems. These pipes should not be concealed before testing, inspecting, and approval.

The Approval

When the proper installation and testing methods are used, approvals come easily.

Chapter 10

Working with Gas

Depending upon the area you work in, your plumbing duties may extend over into working with gas. While gas fitting is not a plumber's job, many plumbers are also gas fitters. Some jurisdictions don't require special licensing for working with gas; others do. Anyone working with gas should be licensed and required to pass a strict examination for the privilege.

Working with gas can be very dangerous. Unlike most plumbing, where a mistake will get you wet, a mistake while working with gas could get you killed. This is not to say that plumbing doesn't present its own set of potentially dangerous circumstances, but the risks of serious injury seem more apparent when working with gas.

Gas work is not usually regulated by the plumbing code, but it is often referred to within the code book. In most jurisdictions the gas code is governed by the mechanical code. Depending upon where you are, plumbing and mechanical codes may overlap. The states in zone one are regulated for gas installations under the plumbing code. This chapter is going to give you a basic understanding of the requirements for working with gas. It is based on a combination of acceptable gas practices in zone one and in other parts of the country. Since zone one is the only major area where gas piping is covered under the plumbing code, there will not be reminder notes for the various regions. Instead special requirements will be identified as they pertain to the states in zone one.

No one without the proper training and experience should work with gas. Even if your jurisdiction does not require a special gas-fitting license, do your homework. Try to learn from your mistakes since your work with gas could be deadly. The risk of personal injury extends beyond the installer. If gas pipe or gas equipment is not installed properly, the people injured or killed as a result of the faulty work could be staggering. When you work with gas, you hold the safety of many people in your hands. Don't take this part of your work lightly.

The two types of gas most often worked with are natural gas and propane gas. There are some differences between them. Equipment that is meant for use with natural gas is not necessarily compatible with propane. Before you make a gas connection to any appliance or equipment, verify the type of gas the unit is intended to work with.

Approved Materials

Several types of piping materials are approved for gas work. All piping used must meet minimum requirements, as established by local codes. The two materials most often used for gas piping in buildings are steel pipe and copper pipe. When copper is used, it should be either type L or type K, and it must be approved for use with gas. Polyvinyl chloride (PVC) and polyethylene (PE) pipe are usually allowed for gas pipe in buried installations outside a building.

Metallic pipe can be used in buildings and above ground so long as the gas being conveyed will not corrode the pipe. Steel pipe, approved copper pipe, and yellow brass pipe are the three types of pipes required for use in zone one. Aluminum pipe, where it is approved, may not be used below ground. It must not be used outside, and when used inside, it may not come into contact with masonry, plaster, or insulation. Further, it must be protected from contact with moisture.

Ductile iron pipe, when approved, is only allowed for underground use, outside of buildings. If any pipe is subject to corrosive action from surrounding conditions, the pipe must be protected to avoid it.

The fittings used with gas pipe must be compatible with the pipe. They must also be approved fittings. When working with gas, bushings are not generally allowed. Increasers and reducers are normally fine. Zone one allows the use of bushings if they are not concealed.

Flexible connectors are often used to connect an appliance to a gas source. These connectors must be approved and marked to prove it. Flex connectors may not be longer than 6 ft. Zone one requires appliance connectors for all appliances, except ranges and dryers, to be no more than 3 ft long. Flex connectors may not be concealed in walls, floors, or partitions. Further, flex connectors may not penetrate walls, floors, or partitions. Flex connectors must be properly sized. They may not be smaller than the inlet of the device they are serving.

Gas hose is not a flex connector. Gas hose is generally prohibited, except for special circumstances. Such circumstances could include a biology lab, where gas burners need to be moved around. If gas hose is approved for use, it must be as short as reasonably possible, and it may not exceed 6 ft in length. This length restriction does not apply to items like hand-held torches.

Gas hose may not be concealed, and it may not penetrate walls, floors, or partitions. If the hose will be exposed to high temperatures, temperatures above 125°F, it may not be used. If allowed for use, gas hose must be connected to a cut-off valve, at the gas pipe supply. This type of hose may be used on outdoor appliances that are designed to be portable. In these uses, the length of the hose may not exceed 15 ft. The hose still must connect to a cut-off valve at the gas pipe supply.

When flex connectors are not used, soft copper tubing often is. In zone one, quick-disconnect connectors are approved. These devices allow the connection to be broken by hand, and the gas is shut off automatically. When copper is used at an appliance connection, it should be type L or type K, and it must not be bent in a manner to damage the structural qualities of the tubing. All pipe bending must be done with approved equipment.

Installing Gas Pipe

Installing gas pipe is not the same as installing plumbing pipes. There are similarities, but the procedure is not the same. One difference is in the way pipe and fittings are put together. All joints must be made gastight. The joints should be tested with a mercury gauge, at the required pressure, to ensure good joints. Zone one requires the test to maintain 6 in of mercury. If tested with a pressure gauge, the test must maintain 10 psi. Air is commonly used to provide the pressure test on gas pipe. The pipe must maintain its test pressure for at least 15 min.

In pipes carrying gas at high pressure, the test pressure is required to be 60 psi. These high-pressure tests are often required to be maintained for 30 min.

All pipe ends are to be cut squarely and with a full diameter. Any burrs on the pipe must be removed. The surfaces of a gas joint must be clean. If flux is used to make a joint, the flux must be approved for the purpose. When installing threaded pipe, only the male threads are allowed to be sealed with pipe dope or tape. Mechanical joints, when used, must be used according to the manufacturer's specification.

The only two types of pipes allowed to have heat-fusion joints are PE and polybutylene (PB), where approved. However, these two types of pipe may not be used with a cement or glue joint. When PVC pipe is used, it must be primed and glued with approved materials.

When more than one type of piping is used, the joint between the opposing pipe types must be made with an approved adapter fitting. In the case of matching metallic pipes together, a dielectric fitting is generally required.

In general, all underground gas piping must be installed at a depth of at least 18 in. Zone one allows buried metallic pipe to be covered by a minimum of 12 in of dirt. The pipe must not be installed in a way to hinder maintenance or to place the pipe in jeopardy of damage. There is, of course, an exception to this rule. Many states allow gas lines serving an individual outside appliance to be buried 8 in deep, but zone one does not recognize this exception. This exception, as usual, is subject to local inspection and approval.

Any underground gas pipe penetrating a foundation must be protected by a pipe sleeve. A rule-of-thumb sizing for the sleeve is two pipe sizes larger than the gas pipe. The additional space in the sleeve must be sealed to prevent water, insect, or vermin invasion. Just as with plumbing, gas pipe located in flood areas must be protected against flooding and the complications associated with flooding.

Piping for gas, other than dry gas, must be graded with a pitch of ¼-in fall for every 15 ft the pipe runs. At any point where the pipe is low or condensation may occur, a drip leg is required. Drip legs must be accessible and must be protected from freezing temperatures. See Fig. 10.1 for a typical drip-leg installation.

The connection of branch piping to a main distribution pipe must be made either on top of the main or on the side but not on the bottom. Bottom connections are not the only prohibited practices. Gas pipe may not be installed in or through heat ducts, air ducts, chimneys, laundry drops, vents, dumbwaiters, or elevator shafts. The reasoning behind this regulation is self-evident. Would you really want your gas pipe running up your chimney?

Concealed piping may not have union connections. Tubing fittings and running threads are also prohibited in concealed locations. There is yet another rule pertaining to concealed gas piping. Unless the pipe is made of steel, it must be protected from punctures. This is most easily accomplished with the use of nail plates. Nail plates are required when a pipe, other than steel, is positioned within 1¼ in from the surface of a wood member, such as a stud or floor joist. The nail plate must have a minimum thickness of ¹⁄₁₆ in. The plate must be large enough to protect the pipe from punctures. Usually, this means installing a plate that extends at least 4 in beyond a normal nailing surface.

When installing gas pipe in concrete, there are still more regulations to observe. Gas pipe buried in concrete must be covered by no less than 1½ in of concrete. The pipe may not make contact with metal objects, and the concrete must not contain materials that will have an adverse effect on the piping.

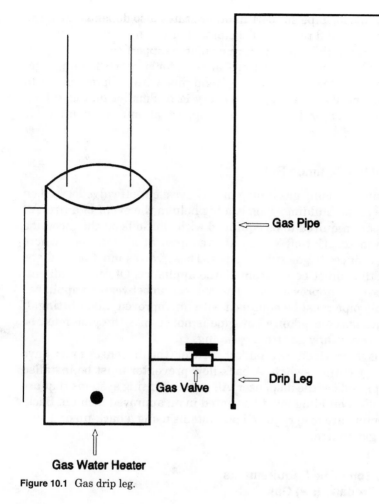

Figure 10.1 Gas drip leg.

Gas-Pipe Supports

Gas pipe can get heavy, and the means of support for the piping must be capable of supporting this weight. Hangers and supports should be made of approved materials that are intended for use with the type of pipe being supported. The required supporting distances are different from plumbing requirements.

All gas pipe installed above ground must be supported in an approved manner and protected from damage. Zone one's requirements for support are as follows: ½-in pipe must be supported at intervals not to exceed 6 ft, ¾- and 1-in pipe at 8-ft intervals, and larger pipe at 10-ft intervals. Pipe with a diameter of at least 1¼ in is required to be supported at each floor level, when installed vertically.

The intervals for pipe support in other states also depends upon the type of pipe used and the size of the pipe. For example, when supporting tubing with a diameter of 1½ in or more, support must be present every 10 ft. This is the same distance allowed for supports holding rigid pipe with a diameter of ¾ in or less. Rigid pipe with a diameter of 1 in or larger only needs to be supported every 12 ft. Smaller tubing, tubing 1¼ in in diameter or less, requires support at minimum intervals of 6 ft.

The Rest of the Common Rules

Every building housing gas piping must have a cut-off valve located on the outside of the building. This is a big help in the event of a fire. All gas meters are required to be equipped with shut-offs on the incoming side of the meter. Cut-off valves are required at all locations where appliances connect to gas supply pipes. These valves must not only be accessible, they must be adjacent to the appliance. Of course, all cut-offs must be of an approved type. The connection between an appliance and a supply pipe must be equipped with an approved union fitting. If an appliance is removed or a gas pipe is not in use, the pipe must be capped to prevent any gas from escaping.

When gas is provided by a bulk dispenser, the dispenser must have an emergency cut-off switch. A back-flow preventer must be installed on the supply side of the dispenser. All gas-dispensing systems that are located inside a building must be vented in an approved fashion. Back-flow preventers are also required on systems using a back-up or a supplemental gas source.

Additional Zone One Requirements
for Liquid Propane (LP) Gas

The relief valves for LP gas must discharge into the open air. These valves must not be located closer than 5 ft, measured horizontally, from any opening into a building. LP gas may not be piped to water heaters in locations where gas might collect and provide opportunity for fire or explosion.

Testing Your Work

All concealed gas piping must be tested and approved before it is concealed. A standard test pressure is a pressure equal to 1½ times the normal working pressure of the system. However, the test pressure must never be less than 3 psi. With LP gas, the test pressure must equate to an 18-in water column. The test must be conducted for a minimum of 10 min. To be approved, the system must not lose pressure during the test. A mercury gauge is the most common way of testing gas pipe.

If the piping loses pressure, leaks should be located with soapy water, not fire or acid. When leaks are located, defective pipe or fittings should be removed and replaced, not repaired.

Once done with the test, the only job left is to purge the system and to get it on line. It is not permissible to purge the gas system through an appliance. The purging must be done in a safe location, where combustion is not a potential threat.

Sizing Gas Pipe

If you thought sizing a potable water system strained your brain, wait until you try sizing gas pipe. No, it's not all that bad, at least not when you use sizing tables. There are formulas available for sizing gas piping, but unless you are a math wizard, they will only serve to frustrate and confuse you.

It is permissible to use a less than exact method when sizing gas pipe. The local gas code will contain tables for your use in selecting the proper pipe sizing. These tables will be based on a few factors that include maximum capacity of the pipe, gas pressure, pressure drop, gravity, and pipe length. The sizing tables provided with gas codes are illustrated and described to make sizing relatively easy.

Most buildings are restricted to a maximum operating pressure of 5 psig. There are exceptions to this rating, but 5 psig is an average rating. This rating is based on the gas in the pipe being natural gas. If the gas is propane, the numbers change; LP gas is meant for a maximum operating pressure of 20 psig. As usual, there will be exceptions to this rule, but 20 psig is normal.

Regulators

Regulators are often needed to regulate gas pressure. If a regulator is used outside, it must be approved for exterior use. Some regulators require an individual vent. When such a vent is required, it must be piped independently to the outside of the building. The vent must be protected against damage and the influx of foreign objects.

Gas regulators must be installed in accessible locations. All regulators must be installed in a manner to prevent them from being damaged. A regulator is required when a gas appliance is designed to work at a lower gas pressure than the pressure present in the piping. If a second-stage regulator is required for LP gas, it must be an approved regulator.

Some More Regulations for Zone One

Zone one has more gas regulations. Used pipe, unless it was used for gas, may not be used in gas installations. If gas pipe is welded, it must be welded by a certified pipeline welder. Exposed gas pipe must be installed in a way to keep it at least 6 in above the ground or other obstructions. It is not permissible to install gas piping below grade within the confines of a building. When special cavities are provided, gas pipe may be concealed and unprotected.

Underground ferrous gas piping must be protected from electricity with isolation fittings that are installed at least 6 in above ground. If unions are installed in gas pipe, they must be installed with right and left nipples and couplings. Unions may be used when they are exposed. When gas pipe serves multiple buildings or tenants, there must be an individual cut-off valve installed for each user. These valves must be installed outside, and they must be readily accessible at all times.

When more than one type of gas has access to a gas pipe, the pipe must be protected against back-flow. Gas-fired barbecues and fireplaces must be controlled with approved valves. The valves must be in the same room as the gas-fired unit. However, the valve may not be in the unit or on a hearth that serves the unit. When installing these valves, they must be installed within 4 ft of the gas outlet for the gas-fired unit. The pipe going from the valve to the unit may be installed in concrete or masonry if the pipe is a standard-weight brass or galvanized steel and there will be at least 2 in of concrete or masonry around the pipe.

Cut-offs for appliances are required to be within 3 ft of the appliance. The cut-offs must be of an approved type. They must be installed on the gas supply pipe. These cut-offs must be installed in front of unions installed between the gas supply pipe and an appliance. These cut-offs can be placed adjacent to, in, or under appliances, so long as the appliance can be moved without affecting the cut-off. When cut-offs are installed in or under gas-fired units, they must be accessible. Appliances may not be piped in a way to allow a gas supply from more than one gas piping system.

When installing underground gas pipe that is not metallic, a number 18 copper wire must also be installed. The wire shall run with and be attached to the gas pipe. The wire must be exposed above grade at both ends of the pipe run.

There you have it; you know the basic requirements for gas piping. However, reading this chapter does not qualify you to work with gas. But if you retain what you have read, you are well on your way to having a working knowledge of gas piping.

Professional Principles

What are professional principles? They are the qualities that separate professional plumbers from renegades. These principles are what make the plumbing profession a trade to be proud of. It is no longer enough to pick up a torch or a pipe wrench and call yourself a plumber. Today, in most areas, a professional plumber must be educated and tested on that education before a plumber's license will be issued. The plumbing codes, the plumbing inspectors, and the plumbers are all a part of what has made plumbing a respectable and profitable career.

For anyone in the plumbing trade, ethics and professional principles are paramount to maintaining the good image plumbers have come to enjoy. Before you have a plumber's license, you may wish licenses were not required. However, once you hold a license, you will be glad the profession is regulated. It is the regulation on the trade that keeps it high on the list of quality incomes. Licensing requirements prohibit just anyone from being a plumber. By requiring all plumbers to be properly licensed, the field of competition is narrowed, and demand for good plumbers remains high. It is this high demand that opens employment opportunities for licensed plumbers and affords them a better-than average income.

The plumbing trade has changed a lot since I entered it. When I started plumbing, I was required to spend my first 6 months in the trade digging ditches and running a jackhammer. Being young, and not very well paid, I considered quitting the trade. It was not until after I buckled down and applied myself that I learned the reason for the hard labor in those first 6 months.

My employer didn't want to spend the time and money training me to be a plumber unless I proved I wanted it badly enough. By sticking with it, I was given the opportunity to work with one of the best plumbers I have ever seen, even to this day. That man, Jerry, trained me to be a real plumber. Jerry didn't cut corners, and he showed me every aspect of the plumbing trade. I learned how to produce quality work in a reasonable time. I learned how to work with inspectors, not against them. In general, Jerry made a very good plumber out of a boy who was not too mechanically inclined at the time.

Times changed and I moved from the foothills of the Blue Ridge Mountains to the big city. When I moved, I was given the position of working foreman. I learned quickly that plumbing wasn't done with the same care and quality in the city that I had learned in the country. I had to adjust my work habits to allow for faster production. While I increased my speed and didn't spend as much time wiping pipe joints and making my work perfect, I refused to do shoddy work.

My personal requirements for doing a professional job cost me a few jobs. Employers were not interested in the quality of the work, only the speed and profit from the job. Being spurred by bad feelings, I tested for my master's license and got it. I was as proud of that license as any other accomplishment I had ever made. I continued to work for other people for some time, but it wasn't long before I opened my own business.

The rest is history. I have been a successful plumbing contractor, along with other ventures, for many years. I attribute much of my personal and financial success to my plumbing endeavors. There have been times when it seemed that if I wasn't a master plumber, I would have gone hungry. But, my plumber's license has never let me down; it has always enabled me to survive and prosper, even through two recessions.

It may seem that I have been rambling about myself, and I have, but for a good reason. I've been involved with plumbing for nearly 20 years, and I know what the trade is all about. For you, the trade may only seem like a job, but it is much more, if you want it to be.

Holding a master plumber's license gives you the opportunity to own your own business, one that is known for its high profits. This opportunity doesn't come without its price. You must dedicate yourself to the trade. You must be willing to pay your dues before you can reap your rewards. If you are looking for a fast track to wealth, plumbing isn't it. But if you are looking for a good career, with unlimited opportunity, plumbing fits the bill.

What does all of this have to do with professional principles? Everything. Professional principles and ethics are the core of any honorable profession. When you assume the role of a plumber, you must bear the responsibilities that go with the position. This book has already taught you many of the rules, regulations, and laws you must follow. This chapter is going to go a little deeper and teach you how to be a better plumber.

Licensing

The first step toward becoming a professional plumber is gaining a license to practice the trade. For many years licenses were not a big deal. Today, however, most jurisdictions require anyone working with plumbing to earn a living to have some type of license. In the old days there were plumber's helpers; today, there are apprentices. By current standards, if you are going to solder joints or glue pipe, you are generally required to have at least an apprentice license.

Digging ditches and carrying pipe can be done without a license, but to become involved enough to learn the trade, you normally need an apprentice license. These licenses are easy to obtain, and they are the first step toward becoming a full-fledged plumber. Generally, all that is required to obtain an apprentice license is the filling out and filing of an application. Apprentice plumbers are not allowed to perform plumbing duties without the direct supervision of either a journeyman or master plumber.

The next license to work toward is a journeyman's license. These licenses are not so easy to come by. Requirements vary, but the requirements for a journeyman license may include educational requirements as well as on-the-job training. The educational requirements can mean spending years in vocational school. The on-the-job training period is usually 4000 h of work. This works out to be about 2 years of full-time work. In most cases, a written test will be administered to determine your knowledge of proper plumbing procedures.

Once educational requirements and on-the-job training are completed, an individual is qualified to take the licensing test. These licensing tests are not easy. In fact, they are often somewhat tricky. The tests usually consist of multiple-choice questions. The degree of difficulty encountered with the test for a journeyman license is much less than that for a master's license, but it is not a task to be taken lightly.

Some areas may waive the requirement for on-the-job training if enough classroom education is received. Likewise, some areas may waive the educational requirements if the amount of on-the-job training is adequate. Students who gain all of their experience in the classroom may not have to take an outside licensing test. Their tests may be administered in the classroom apprenticeship program.

Once a person becomes a journeyman plumber, that individual is allowed to perform plumbing without direct supervision. Journeyman plumbers are generally not allowed to go into business for themselves unless they employ a master plumber. Most areas require journeyman plumbers to work under the supervision of a master plumber but not direct supervision. This basically means that a journeyman plumber can go out on a job and work alone as long as a master plumber reviews the work from time to time.

After becoming licensed as a journeyman plumber, an individual is usually required to work for a period of time before sitting for the master plumber's test. This period of time is usually 2000 h, or about 1 year of full-time work. Once the time requirements are met, an individual can take the test for a master's license. These tests are the most difficult of all.

Like the journeyman's test, the master's test also contains numerous multiple-choice questions. However, the type and complexity of these questions are more demanding than those for a journeyman's license. A master's applicant will generally have to perform some type of pipe sizing for plumbing and storm drainage. Master licenses are only awarded to those with a thorough knowledge of the plumbing code. Many first-time applicants are unsuccessful in their quest for a passing grade.

To improve the odds of passing either the journeyman or master test, individuals can attend test-preparation courses. These courses are designed to hone the knowledge of students and to prepare them for the licensing exams. Normally, these courses are offered in the evenings. The length of the course varies, but they usually are held one or two nights a week for several weeks.

If students pay attention, these classes offer a big advantage. The instructor usually has a good feel for the type of material the students will be tested on for their licensing requirements. It is standard procedure in these exam preparation courses for the instructor to give many tests during the course. The tests are often similar to the ones encountered when taking a licensing exam. By taking these sample tests, the students learn many aspects of passing the real test.

The students learn what areas they are deficient in. They learn how to study, and this is a big factor in passing any test. Instructors help clear confusion students may have in interpreting the local plumbing code. By the end of a code-awareness class, students have a significant edge during their licensing exams.

One of the most important aspects of passing a licensing exam is understanding the plumbing code. The way licensing tests are often worded, simply memorizing the code book is not enough to pass the tests. To achieve good scores, applicants must be able to interpret and apply the code. This is often much more difficult than simply being able to quote the code in chapter and verse.

After the licensing tests are taken and passed, a license will be forthcoming. It is very important to pay particular attention to the answer you give on the application for a license. By not providing a truthful account of your past, you may be rejected from consideration for licensing. If a license is issued based on false or incomplete application information, the license may be revoked. Something as simple as forgetting traffic violations can cause trouble with your application. Answer all questions, and answer them honestly.

Once the license is issued, you normally are required to display it in the place of business where you will work. This generally holds true for apprentices, journeymen, and masters. Most areas also issue a small, pocket-sized license along with the wall certificate. These pocket cards are generally required to be carried at all times when plumbing is being performed. The cards allow a means for a code enforcement officer to verify the license status of all people working on a job.

Once you have your license, don't do anything to risk losing it. If a permit is required for the work you will be performing, obtain a permit before starting the work. Do all work in compliance with the local code requirements. Never try to fool an inspector with tricks. Most inspectors know the tricks and will catch you. Even if you don't lose your license, you will lose the respect of the code officer. Consequently, all of your future work will be inspected with a fine-toothed comb.

Abide by all code requirements. Some jurisdictions require a plumber's license number to be listed in all advertisements and on all work vehicles. If this is a requirement in your area, comply with it. Some jurisdictions require continuing education for licensed plumbers. If you want to keep your license, you must attend these seminars. You will be doing yourself a favor to learn from the seminars rather than simply attending them.

For your own benefit, join plumbing organizations and read; read a lot. The more you learn, the better prepared you will be to make a better living. Don't argue with code officers. If you feel they are wrong, approach the subject but do so diplomatically. Remember, much of the plumbing code is left to the interpretation of the local code officer.

Avoid doing sloppy work. Not only will inspectors check your work more closely when it is shoddy, you will not be in as much demand as a plumber. Quality work pays off in many ways. When a job is done right, you eliminate callbacks. Having to go back and fix improper work gets expensive. If you have your own business, it will not take long to see the advantages of doing the job right the first time. If you are an employee who is frequently creating callbacks, you may not be employed for long.

Good work will result in referral business. There is no better source of new business than referrals. By giving customers their money's worth, you are building business for yourself and your employer. Even if you don't own the company, if the company doesn't have ongoing work, neither do you.

Take pride in your chosen field, and don't let the new image of plumbers be tarnished. It takes a multitude of professional plumbers to build a strong image, but it only takes a few bad plumbers to ruin it.

Lastly, take the time to train new plumbers. Remember, you wouldn't have your present opportunities if someone had not taken the time to train you properly. With rising costs and the desire to cut expenses, too many plumbers don't spend the time to train new help. This creates a void in the trade. It is also a fact that today's worker is not as interested in learning the trade as in making money. Many young people don't understand that they must learn before they can earn.

In closing, remember this—plumbers protect the health of the nation. Without plumbers, our world would be a much worse place to live, work, and play. Today, more than ever, water conservation, pollution control, and sanitary conditions are primary concerns. It is up to professional plumbers to see that these needs are met and improved upon.

Chapter

12

Charts and Tables

This chapter is filled with charts, tables, and other illustrations. These illustrations are similar to those found in plumbing code books. By looking over these illustrations, you can get a good feel for how a code book provides information for you to work with. Review the illustrations to learn what to look for in your code book. Note that the cross-references refer to other tables in the code book that may not be reproduced here.

Type of fixture or device	Pipe size (in)	Type of fixture or device	Pipe size (in)
Bath Tubs	½	Shower (Single Head)	½
Combination Sink		Sinks (Serv., Slop)	½
and Tray	½	Sinks Flushing Rim	¾
Drinking Fountain	⅜	Urinal (Flush Tank)	½
Dishwasher		Urinal (Direct	
(Domestic)	½	Flush Valve)	¾
Kitchen Sink, Residential	½	Water Closet	
Kitchen Sink, Commercial	¾	(Flushometer Valve Type)	1
Lavatory	⅜	Water Closet	
Laundry Tray, 1, 2, or 3		(Gravity or Flushometer	
Compartments	½	Tank Type)	⅜
Wall Hydrants	½	Hose Bibbs	½

Figure 12.1 Size of fixture supply, Table 1211.3.1 of the code. (*Reproduced from the 1991 edition of* The Standard Plumbing Code® *with permission of the copyright holder, Southern Building Code Congress, International, Inc. All rights reserved.*)

Supply systems predominantly for flush tanks			Supply systems predominantly for flush valves		
Load	Demand		Load	Demand	
(Water supply fixture units)	(Gallons per minute)	(Cubic feet per minute)	(Water supply fixture units)	(Gallons per minute)	(Cubic feet per minute)
1	3.0	0.04104			
2	5.0	0.0684			
3	6.5	0.86892			
4	8.0	1.06944			
5	9.4	1.256592	5	15.0	2.0052
6	10.7	1.430376	6	17.4	2.326032
7	11.8	1.577424	7	19.8	2.646364
8	12.8	1.711104	8	22.2	2.967696
9	13.7	1.831416	9	24.6	3.288528
10	14.6	1.951728	10	27.0	3.60936
11	15.4	2.058672	11	27.8	3.716304
12	16.0	2.13888	12	28.6	3.823248
13	16.5	2.20572	13	29.4	3.930192
14	17.0	2.27256	14	30.2	4.037136
15	17.5	2.3394	15	31.0	4.14408
16	18.0	2.90624	16	31.8	4.241024
17	18.4	2.459712	17	32.6	4.357968
18	18.8	2.513184	18	33.4	4.464912
19	19.2	2.566656	19	34.2	4.571856
20	19.6	2.620128	20	35.0	4.6788
25	21.5	2.87412	25	38.0	5.07984
30	23.3	3.114744	30	42.0	5.61356
35	24.9	3.328632	35	44.0	5.88192
40	26.3	3.515784	40	46.0	6.14928
45	27.7	3.702936	45	48.0	6.41664
50	29.1	3.890088	50	50.0	6.684
60	32.0	4.27776	60	54.0	7.21872
70	35.0	4.6788	70	58.0	7.75344
80	38.0	5.07984	80	61.2	8.181216
90	41.0	5.48088	90	64.3	8.595624
100	43.5	5.81508	100	67.5	9.0234
120	48.0	6.41664	120	73.0	9.75864
140	52.5	7.0182	140	77.0	10.29336
160	57.0	7.61976	160	81.0	10.82808
180	61.0	8.15448	180	85.5	11.42964
200	65.0	8.6892	200	90.0	12.0312
225	70.0	9.3576	225	95.5	12.76644
250	75.0	10.0260	250	101.0	13.50168
275	80.0	10.6944	275	104.5	13.96956
300	85.0	11.3628	300	108.0	14.43744
400	105.0	14.0364	400	127.0	16.97736
500	124.0	16.57632	500	143.0	19.11624
750	170.0	22.7256	750	177.0	23.66136
1000	208.0	27.80544	1000	208.0	27.80544
1250	239.0	31.94952	1250	239.0	31.94952
1500	269.0	35.95992	1500	269.0	35.95992
1750	297.0	39.70296	1750	297.0	39.70296
2000	325.0	43.446	2000	325.0	43.446
2500	380.0	50.7984	2500	380.0	50.7984
3000	433.0	57.88344	3000	433.0	57.88344
4000	535.0	70.182	4000	525.0	70.182
5000	593.0	79.27224	5000	593.0	79.27224

Figure 12.2 Table for estimating demand, Table F102 of the code. (*Reproduced from the 1991 edition of* The Standard Plumbing Code® *with permission of the copyright holder, Southern Building Code Congress, International, Inc. All rights reserved.*)

Fixture	Occupancy	Type of supply control	Load values, in water supply fixture units		
			Cold	Hot	Total
Bathroom group	Private	Flush tank	4.5	3.0	6.0
Bathroom group	Private	Flush valve	6.0	3.0	8.0
Bathtub	Private	Faucet	1.5	1.5	2.0
Bathtub	Public	Faucet	3.0	3.0	4.0
Bidet	Private	Faucet	1.5	1.5	2.0
Combination fixture	Private	Faucet	2.25	2.25	3.0
Dishwashing machine	Private	Automatic		1.0	1.0
Drinking fountain	Offices, etc.	⅜″ valve	0.25		0.25
Kitchen sink	Private	Faucet	1.5	1.5	2.0
Kitchen sink	Hotel, Restaurant	Faucet	3.0	3.0	4.0
Laundry trays (1 to 3)	Private	Faucet	2.25	2.25	3.0
Lavatory	Private	Faucet	0.75	0.75	1.0
Lavatory	Public	Faucet	1.5	1.5	2.0
Service sink	Offices, etc.	Faucet	2.25	2.25	3.0
Shower head	Public	Mixing valve	3.0	3.0	4.0
Shower stall	Private	Mixing valve	1.5	1.5	2.0
Urinal	Public	1″ flush valve	10.0		10.0
Urinal	Public	¾″ flush valve	5.0		5.0
Urinal	Public	Flush tank	3.0		3.0
Washing machine (8 lb)	Private	Automatic	1.5	1.5	3.0
Washing machine (8 lb)	Public	Automatic	2.25	2.25	2.0
Washing machine (16 lb)	Public	Automatic	3.0	3.0	3.0
Water closet	Private	Flush valve	6.0		4.0
Water closet	Private	Flush tank	3.0		6.0
Water closet	Public	Flush valve	10.0		3.0
Water closet	Public	Flush tank	5.0		10.0
Water closet	Public or Private	Flushometer Tank	2.0		5.0
					2.0

NOTE: For fixtures not listed, loads should be assumed by comparing the fixture to one listed using water in similar quantities and at similar rates. The assigned loads for fixtures with both hot and cold water supplies are given for separate hot and cold water loads and for total load, the separate hot and cold water loads being three-fourths of the total load for the fixture in each case.

Figure 12.3 Load values assigned to fixtures, Table F101 of the code. *(Reproduced from the 1991 edition of The Standard Plumbing Code® with permission of the copyright holder, Southern Building Code Congress, International, Inc. All rights reserved.)*

Required leaching area (sq ft/100 gal septic) tank capacity	Maximum septic tank size allowable
20–25	7500
40	5000
60	3500
90	3000

Figure 12.4 Septic tank size, Table E102C of the code. (*Reproduced from the 1991 edition of* The Standard Plumbing Code® *with permission of the copyright holder, Southern Building Code Congress, International, Inc. All rights reserved.*)

Type of soil	Req'd leaching area (sq ft/100 gal)	Maximum absorption capacity gals/sq ft of leaching area for a 24-hour period
1. Coarse sand or gravel	20	5
2. Fine sand	25	4
3. Sandy loam or sandy clay	40	2.5
4. Clay with considerable sand or gravel	60	1.66
5. Clay with small amount of sand or gravel	90	1.11

Figure 12.5 Rated absorption capacities of five typical soils, Table E103 of the code. (*Reproduced from the 1991 edition of* The Standard Plumbing Code® *with permission of the copyright holder, Southern Building Code Congress, International, Inc. All rights reserved.*)

Type of building	Daily per capita[1] (gal)	Basic factor
Grammar School	15	per classroom
Grammar School with Cafeteria	20	(35 students per classroom)
High School with Cafeteria and Shower Baths	25	
Factories	20 (without showers)	Each 8-hour shift
	25 (with showers)	Each 8-hour shift
Restaurants	50	Per seat
Trailer Parks—Community Baths	50	3 persons per trailer
Trailer Parks—Private Baths or Independent Trailers	60	3 persons per trailer
Motels—Baths and Toilets	50	3 persons per unit
Motels—Bath, Toilet, and Kitchen	60	3 persons per unit
Self-Service Laundry	300	per machine
Drive-in Theaters	5	per car

[1] Normal sludge storage capacity is included excepting waste from food disposal units.

Figure 12.6 Septic tank capacity, Table E102B of the code. (*Reproduced from the 1991 edition of* The Standard Plumbing Code® *with permission of the copyright holder, Southern Building Code Congress, International, Inc. All rights reserved.*)

Single family dwellings[2]—number of bedrooms	Multiple dwelling units[3] or apartments—one bedroom each	Other uses: maximum fixture units[4] served per Table 922.2	Minimum septic tank capacity in gallons
1 or 2		15	750
3		20	1000
4	2 units	25	1200
5 or 6	3	33	1500
	4	45	2000
	5	55	2250
	6	60	2500
	7	70	2750
	8	80	3000
	9	90	3250
	10	100	3500

[1] Septic tank sizes in this table include sludge storage capacity and the connection of domestic food waste disposal units without further volume increase.
[2] Extra bedroom, 150 gal each.
[3] Extra dwelling units over 10, 250 gal each.
[4] Extra fixture units over 100, 25 gal per fixture unit.

Figure 12.7 Capacity of septic tanks, Table E102A of the code.[1] (*Reproduced from the 1991 edition of* The Standard Plumbing Code® *with permission of the copyright holder, Southern Building Code Congress, International, Inc. All rights reserved.*)

Distance from:	Building sewer	Septic tank	Disposal field	Seepage pit or cesspool
Buildings or structures[1]	2	5	8	8
Property line adjoining private property	Clear	5	5	8
Water supply wells	50	50	100	150
Streams	50	50	50	100
Large trees			10	10
Seepage pits or cesspools		5	5	12
Disposal field		5	4[3]	5
Domestic water line	1	5	5	5
Distribution box		5	5	5

[1] Including porches and steps whether covered or uncovered, breezeways, roofed porte-cocheres, roofed patios, carports, covered walks, covered driveways and similar structures or appurtenances.
[2] All nonmetallic drainage piping shall clear domestic water supply wells by at least 50 ft. This distance may be reduced to not less than 25 ft when approved type metallic piping is installed. Where special hazards are involved the distance required shall be increased, as may be directed by the Health Officer or the Plumbing Official.
[3] Plus 2 ft for each additional foot of depth in excess of 1 ft below the bottom of the drain line. (See also E106.)
[4] When disposal fields and/or seepage pits are installed in sloping ground the minimum horizontal distance between any part of the leaching system and ground surface shall be 15 ft.

Figure 12.8 Location of sewage disposal system, Table E101 of the code.[4] (*Reproduced from the 1991 edition of* The Standard Plumbing Code® *with permission of the copyright holder, Southern Building Code Congress, International, Inc. All rights reserved.*)

Component	Design recommendations	Possible troubles to be guarded against
Bend at foot of stack	Bend to be "large radius" i.e. 6-in minimum root radius or, if adequate vertical distance is available, two "large radius" 45° bends are to be preferred. Vertical distance between lowest branch connection and invert of drain to be at least 18 in for a two story house and 30 in for taller dwellings. Where this distance cannot be achieved, ground floor fixtures shall be connected directly to the building drain and vented as provided for in other chapters of this code. See Figure 1602.4B.	Back pressure at lowest branch, foaming of detergents
W.C. branch connection to stack	Water closet connections shall be swept in the direction of flow with radius at the invert of not less than 2 in. Fittings in other materials shall have the same sweep as cast iron fittings. The length of unvented water closet branches shall be limited by the diameter of the branch piping: 6 ft for 3-in dia., 10 ft for 4-in dia.	Induced siphonage at lower level in the stack when water closets are discharged.
Lavatory waste 1¼-in trap and 1½-in minimum waste pipe Lavatories with 1½-in P.O. plugs may be installed as provided for sink waste	"P" traps shall be used. The maximum fall of the waste pipe shall not exceed the hydraulic gradient of the pipe. For the maximum distance between the stack and trap weir see Table 1602.5B. Any bends on plan shall be of not less than 3-in radius at the center line. Waste pipes longer than the recommended maximum length shall be vented. As an alternative, 2-in diameter waste pipes may be used so long as the hydraulic gradient is not exceeded, but additional maintenance may be necessary to maintain the bore.	Self-siphonage

[1] Where the length or fall of the discharge pipe serving a waste fixture is greater than the recommended maximum in this table, the discharge pipe shall preferably be vented (see 1602.6) or a larger diameter discharge pipe shall be used. This may have a maximum length of 10 ft.

Figure 12.9 Design of single branches and fittings, Table 1602.5A of the code.[1] (*Reproduced from the 1991 edition of* The Standard Plumbing Code® *with permission of the copyright holder, Southern Building Code Congress, International, Inc. All rights reserved.*)

Component	Design recommendations	Possible troubles to be guarded against
Bath waste 1½-in trap and 1½-in waste pipe	"P" traps shall be used (a 2-in parallel branch, when required, shall not be considered a violation of requirements of other sections of this Code, when its vertical length does not exceed 12½ in, and the center line of the parallel branch is not more than 12½ in from the stack). Owing to the flat bottom of a bath, the trailing discharge normally refills the trap and the risk of self-siphonage is much reduced. Waste pipes 7 ft 6 in long at a fall of ¼ in/ft have been used successfully. Position of entry of bath waste into stack to be as shown in Figure 1602.4A.	Self-siphonage
Sink waste 1½-in trap and 1½-in waste pipe	"P" traps shall be used. Owing to the flat bottom of a sink the trailing discharge normally refills the trap and the risk of self-siphonage is much reduced. Fall of ¼ in/ft shall be maintained. For maximum length see Table 1602.5B. A sink with 1½-in tail piece may be drained with 2-in horizontal branch not exceeding 8 feet in length. When a 2-in branch is used, the trap outlet shall connect to a 2-in × 1½-in reducing fitting. An opening into the branch larger than 1½ in will not be permitted except for a cleanout.	Self-siphonage

[1] Where the length or fall of the discharge pipe serving a waste fixture is greater than the recommended maximum in this table, the discharge pipe shall preferably be vented (see 1602.6) or a larger diameter discharge pipe shall be used. This may have a maximum length of 10 ft.

Figure 12.9 (*Continued*)

Size of fixture[1] drain (in)	Distance trap to stack or vent
Note 4	4 ft 6 in
1½	5 ft 6 in[2]
2	7 ft 6 in[3]

[1] Minimum size.
[2] Other than bath waste.
[3] For bathtubs see Table 1602.5A.
[4] 1½-inch fixture drains required for lavatories with 1¼-inch traps.

Figure 12.10 Distance from trap weir to stack or other ventilating pipe, Table 1602.5B of the code. (*Reproduced from the 1991 edition of* The Standard Plumbing Code® *with permission of the copyright holder, Southern Building Code Congress, International, Inc. All rights reserved.*)

Nominal internal diameter of pipe (in)	Fall per foot (in)		
	⅛	¼	½
2		10	26
2½		35	95
3	40	100	230
4	230	430	1050
5	780	1500	3000
6	2000	3500	7500

[1] Discharge pipes sized by this method give the minimum size necessary to carry the expected flow load. Separate ventilation pipes may be required (see 1602.6). It may be worthwhile to consider oversizing the discharge pipes to reduce the ventilating pipework required.
[2] Building sewer sizes start at 4 inches.

Figure 12.11 Maximum number of discharge units connected to building drain or building sewer, Table 1602.4C of the code.[1,2] (*Reproduced from the 1991 edition of* The Standard Plumbing Code® *with permission of the copyright holder, Southern Building Code Congress, International, Inc. All rights reserved.*)

Type of domestic appliance	Min. internal diameter (in)
Lavatory	1¼
Sink	1½
Bathtub	1½ or 2
Shower	2
Wash tub	1½
Kitchen waste disposal unit (tubular trap is essential)	1½

Figure 12.12 Minimum internal diameters of traps, Table 1602.3 of the code. (*Reproduced from the 1991 edition of* The Standard Plumbing Code® *with permission of the copyright holder, Southern Building Code Congress, International, Inc. All rights reserved.*)

Size of leader or conductor[1] (in)	Maximum projected roof area (sq ft)
2	720
2½	1300
3	2200
4	4600
5	8650
6	13,500
8	29,000

[1] The equivalent diameter of square or rectangular leader may be taken as the diameter of that circle which may be inscribed within the cross-sectional area of the leader. See 1506.2.2.

Figure 12.13 Size of vertical leaders, Table 1506.1 of the code. (*Reproduced from the 1991 edition of* The Standard Plumbing Code® *with permission of the copyright holder, Southern Building Code Congress, International, Inc. All rights reserved.*)

Diameter of drain (in)	Maximum projected roof area for drains of various slopes (sq ft)		
	⅛ in slope	¼ in slope	½ in slope
3	822	1160	1644
4	1880	2650	3760
5	3340	4720	6680
6	5350	7550	10,700
8	11,500	16,300	23,000
10	20,700	29,200	41,400
12	33,300	47,000	66,600
15	59,500	84,000	119,000

Figure 12.14 Size of horizontal storm drains, Table 1506.2 of the code. (*Reproduced from the 1991 edition of* The Standard Plumbing Code® *with permission of the copyright holder, Southern Building Code Congress, International, Inc. All rights reserved.*)

Soil or waste pipe diam. (in)	Fixture units (max. no.)	Diameter of circuit or loop vent (in)					
		1½	2	2½	3	4	5
2	3	15	40				
2½	6	10	30				
3	10		20	40	100		
4	80		7	20	52	200	
5	180				16	70	200

Figure 12.15 Battery vent sizing table, Table 1420.3 of the code. (*Reproduced from the 1991 edition of* The Standard Plumbing Code® *with permission of the copyright holder, Southern Building Code Congress, International, Inc. All rights reserved.*)

Diameter horizontal drainage branch (in)	Slope of drainage branch (in/ft)	Diameter of vent (in)							
		1¼	1½	2	2½	3	4	5	6
1¼	¼	*							
1½	¼	*	*						
2	¼	*	*	*					
2½	¼	805	*	*	*				
3	⅛	660	*	*	*	*			
3	¼	335	710	*	*	*			
4	⅛	132	364	*	*	*	*		
4	¼	69	147	600	*	*	*		
5	⅛		94	355	845	*	*	*	
5	¼		47	160	412	*	*	*	
6	½		35	124	320	*	*	*	*
6	¼			60	155	*	*	*	*

* More than 1000 ft.

Figure 12.16 Individual and branch vent sizing table, Table 1420.4 of the code. (*Reproduced from the 1991 edition of* The Standard Plumbing Code® *with permission of the copyright holder, Southern Building Code Congress, International, Inc. All rights reserved.*)

Size of soil or waste stack (in)	Fixture units connected	Diameter of vent required (in)								
		1¼	1½	2	2½	3	4	5	6	8
1¼	2	30								
1½	8	50	150							
1½	10	30	100							
2	12	30	75	200						
2	20	26	50	150						
2½	42		30	100	300					
3	10		30	100	200	600				
3	30			60	200	500				
3	60			50	80	400				
4	100			35	100	260	1000			
4	200			30	90	250	900			
4	500			20	70	180	700			
5	200				35	80	350	1000		
5	500				30	70	300	900		
5	1100				20	50	200	700		
6	350				25	50	200	400	1300	
6	620				15	30	125	300	1100	
6	960					24	100	250	1000	
6	1900					20	70	200	700	
8	600						50	150	500	1300
8	1400						40	100	400	1200
8	2200						30	80	350	1100
8	3600						25	60	250	800
10	1000							75	125	1000
10	2500							50	100	500
10	3800							30	80	350
10	5600							25	60	250

Figure 12.17 Maximum length of stack vents, vent stacks, and relief vents, Table 1420.2 of the code. (*Reproduced from the 1991 edition of* The Standard Plumbing Code® *with permission of the copyright holder, Southern Building Code Congress, International, Inc. All rights reserved.*)

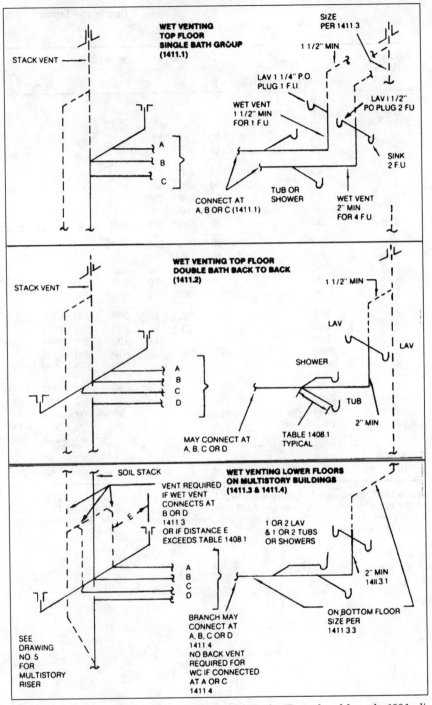

Figure 12.18 Wet venting examples, Figure 10 of the code. (*Reproduced from the 1991 edition of* The Standard Plumbing Code® *with permission of the copyright holder, Southern Building Code Congress, International, Inc. All rights reserved.*)

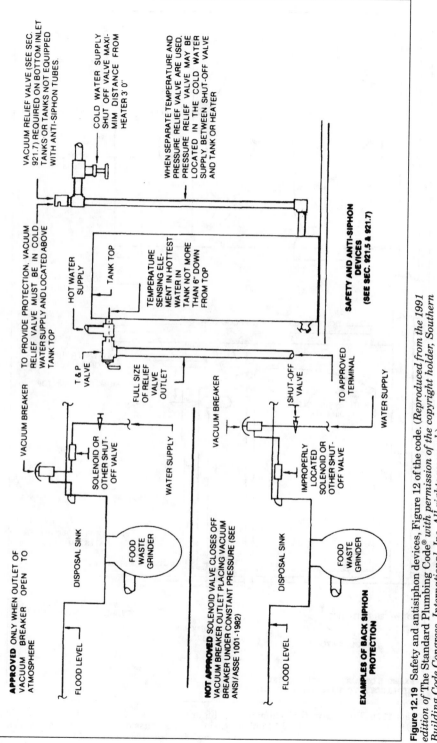

Figure 12.19 Safety and antisiphon devices, Figure 12 of the code. (*Reproduced from the 1991 edition of* The Standard Plumbing Code® *with permission of the copyright holder, Southern Building Code Congress, International, Inc. All rights reserved.*)

227

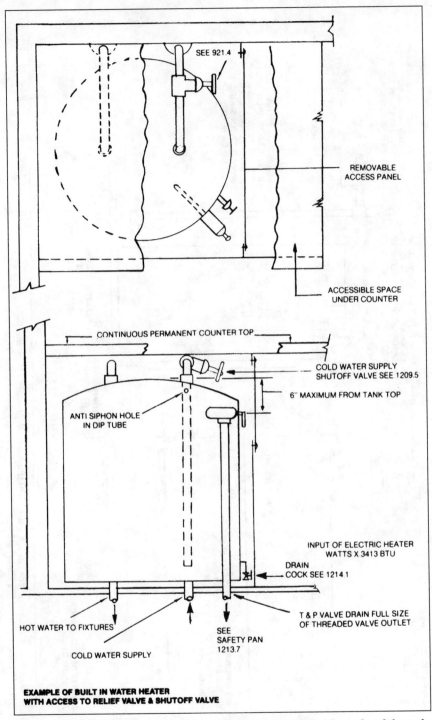

SEE 921.4

REMOVABLE ACCESS PANEL

ACCESSIBLE SPACE UNDER COUNTER

CONTINUOUS PERMANENT COUNTER TOP

COLD WATER SUPPLY SHUTOFF VALVE SEE 1209.5

6" MAXIMUM FROM TANK TOP

ANTI SIPHON HOLE IN DIP TUBE

INPUT OF ELECTRIC HEATER WATTS X 3413 BTU

DRAIN COCK SEE 1214.1

T & P VALVE DRAIN FULL SIZE OF THREADED VALVE OUTLET

HOT WATER TO FIXTURES

SEE SAFETY PAN 1213.7

COLD WATER SUPPLY

EXAMPLE OF BUILT IN WATER HEATER WITH ACCESS TO RELIEF VALVE & SHUTOFF VALVE

Figure 12.20 Built-in water heater access, Figure 9 of the code. (*Reproduced from the 1991 edition of* The Standard Plumbing Code® *with permission of the copyright holder, Southern Building Code Congress, International, Inc. All rights reserved.*)

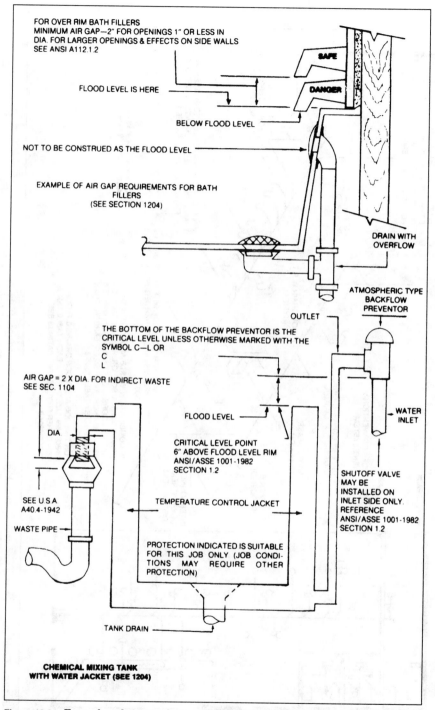

FOR OVER RIM BATH FILLERS
MINIMUM AIR GAP—2" FOR OPENINGS 1" OR LESS IN
DIA. FOR LARGER OPENINGS & EFFECTS ON SIDE WALLS
SEE ANSI A112.1.2

SAFE

DANGER

FLOOD LEVEL IS HERE

BELOW FLOOD LEVEL

NOT TO BE CONSTRUED AS THE FLOOD LEVEL

EXAMPLE OF AIR GAP REQUIREMENTS FOR BATH
FILLERS
(SEE SECTION 1204)

DRAIN WITH
OVERFLOW

ATMOSPHERIC TYPE
BACKFLOW
PREVENTOR

OUTLET

THE BOTTOM OF THE BACKFLOW PREVENTOR IS THE
CRITICAL LEVEL UNLESS OTHERWISE MARKED WITH THE
SYMBOL C—L OR
C
L

AIR GAP = 2 X DIA. FOR INDIRECT WASTE
SEE SEC. 1104

WATER
INLET

DIA.

FLOOD LEVEL

CRITICAL LEVEL POINT
6" ABOVE FLOOD LEVEL RIM
ANSI/ASSE 1001-1982
SECTION 1.2

SEE U.S.A.
A40.4-1942

TEMPERATURE CONTROL JACKET

SHUTOFF VALVE
MAY BE
INSTALLED ON
INLET SIDE ONLY.
REFERENCE
ANSI/ASSE 1001-1982
SECTION 1.2

WASTE PIPE

PROTECTION INDICATED IS SUITABLE
FOR THIS JOB ONLY (JOB CONDI-
TIONS MAY REQUIRE OTHER
PROTECTION)

TANK DRAIN

CHEMICAL MIXING TANK
WITH WATER JACKET (SEE 1204)

Figure 12.21 Examples of air gap distances, Figure 8 of the code. (*Reproduced from the 1991 edition of* The Standard Plumbing Code® *with permission of the copyright holder, Southern Building Code Congress, International, Inc. All rights reserved.*)

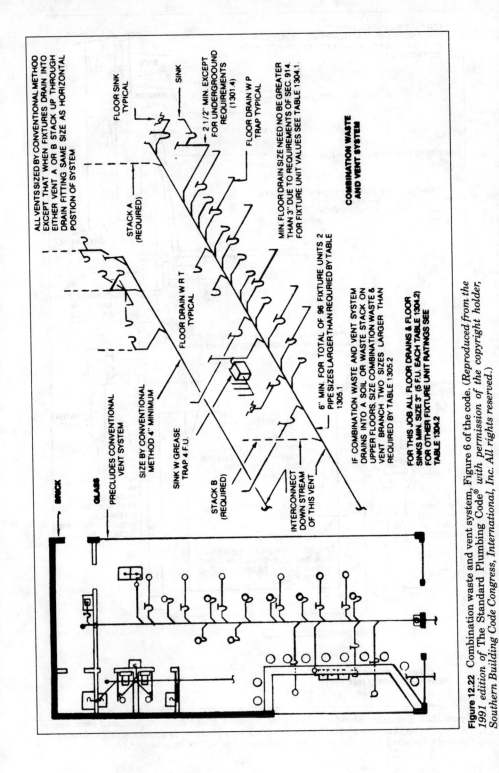

Figure 12.22 Combination waste and vent system, Figure 6 of the code. (*Reproduced from the 1991 edition of The Standard Plumbing Code® with permission of the copyright holder, Southern Building Code Congress, International, Inc. All rights reserved.*)

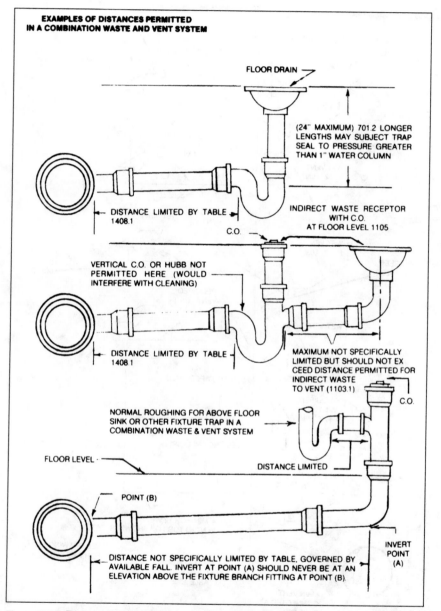

EXAMPLES OF DISTANCES PERMITTED IN A COMBINATION WASTE AND VENT SYSTEM

FLOOR DRAIN

(24" MAXIMUM) 701.2 LONGER LENGTHS MAY SUBJECT TRAP SEAL TO PRESSURE GREATER THAN 1" WATER COLUMN

DISTANCE LIMITED BY TABLE 1408.1

C.O.

INDIRECT WASTE RECEPTOR WITH C.O. AT FLOOR LEVEL 1105

VERTICAL C.O. OR HUBB NOT PERMITTED HERE (WOULD INTERFERE WITH CLEANING)

DISTANCE LIMITED BY TABLE 1408.1

MAXIMUM NOT SPECIFICALLY LIMITED BUT SHOULD NOT EXCEED DISTANCE PERMITTED FOR INDIRECT WASTE TO VENT (1103.1)

C.O.

NORMAL ROUGHING FOR ABOVE FLOOR SINK OR OTHER FIXTURE TRAP IN A COMBINATION WASTE & VENT SYSTEM

FLOOR LEVEL

DISTANCE LIMITED

POINT (B)

INVERT POINT (A)

DISTANCE NOT SPECIFICALLY LIMITED BY TABLE, GOVERNED BY AVAILABLE FALL. INVERT AT POINT (A) SHOULD NEVER BE AT AN ELEVATION ABOVE THE FIXTURE BRANCH FITTING AT POINT (B).

Figure 12.23 Distances permitted in a combination waste and vent system, Figure 7 of the code. (*Reproduced from the 1991 edition of* The Standard Plumbing Code® *with permission of the copyright holder, Southern Building Code Congress, International, Inc. All rights reserved.*)

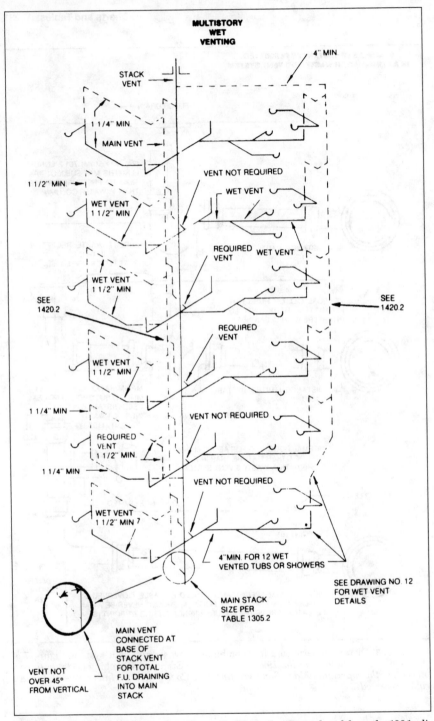

Figure 12.24 Multistory wet venting, Figure 5 of the code. (*Reproduced from the 1991 edition of* The Standard Plumbing Code® *with permission of the copyright holder, Southern Building Code Congress, International, Inc. All rights reserved.*)

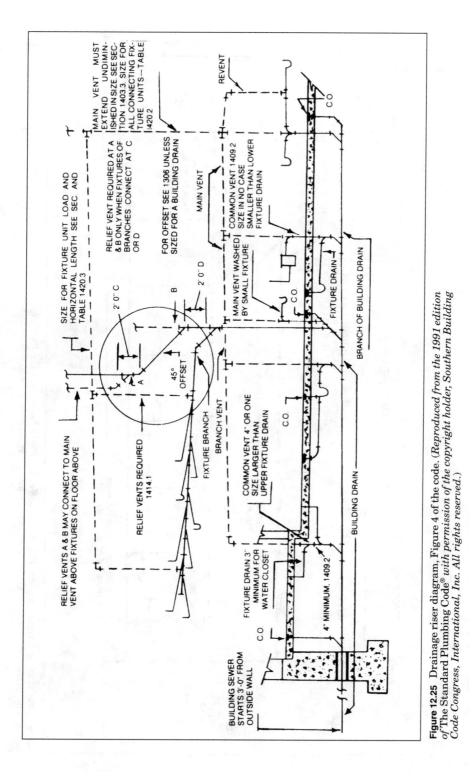

Figure 12.25 Drainage riser diagram, Figure 4 of the code. (*Reproduced from the 1991 edition of* The Standard Plumbing Code® *with permission of the copyright holder, Southern Building Code Congress, International, Inc. All rights reserved.*)

233

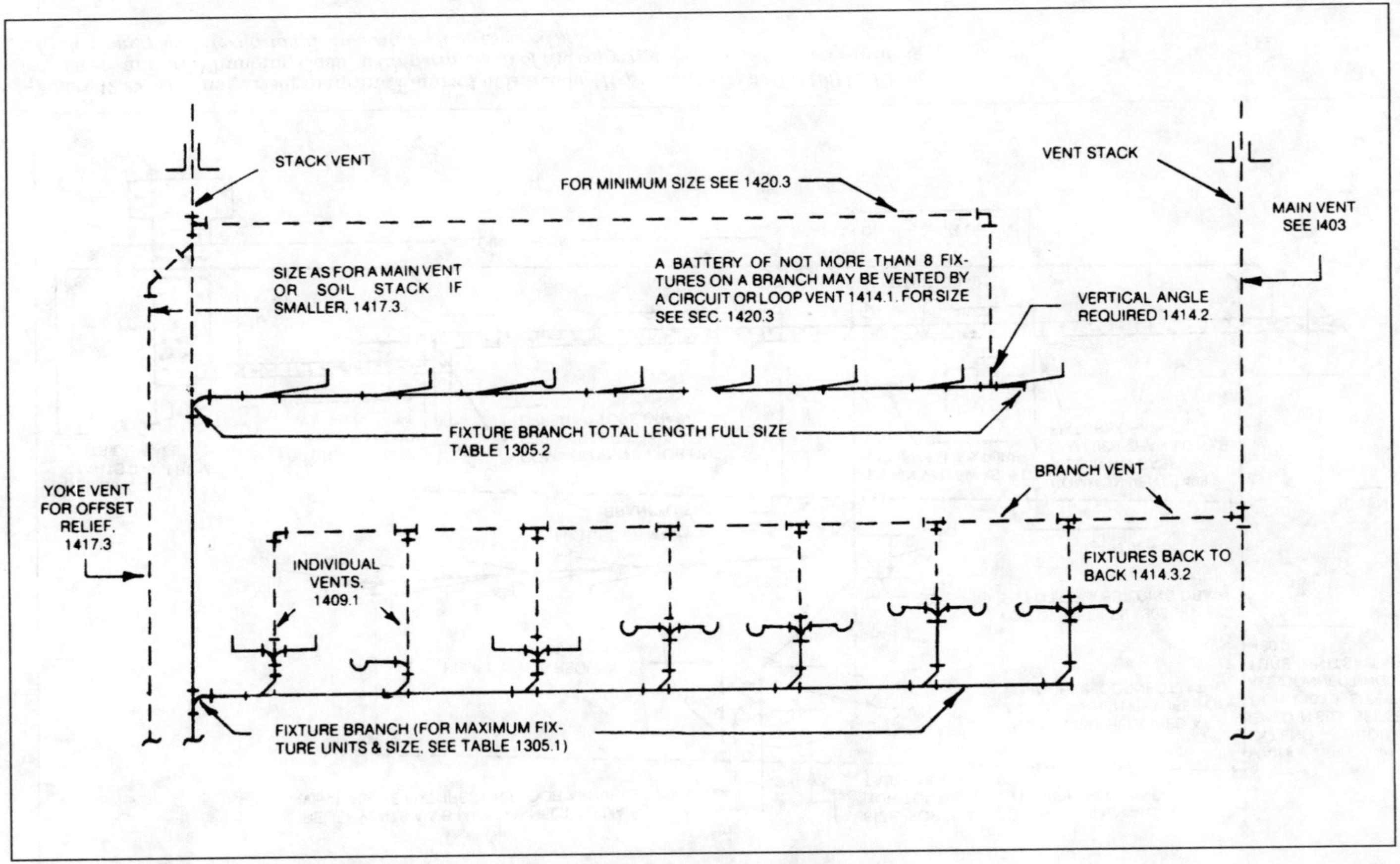
STACK VENT
FOR MINIMUM SIZE SEE 1420.3
VENT STACK
MAIN VENT SEE I403
SIZE AS FOR A MAIN VENT OR SOIL STACK IF SMALLER, 1417.3.
A BATTERY OF NOT MORE THAN 8 FIXTURES ON A BRANCH MAY BE VENTED BY A CIRCUIT OR LOOP VENT 1414.1. FOR SIZE SEE SEC. 1420.3
VERTICAL ANGLE REQUIRED 1414.2.
FIXTURE BRANCH TOTAL LENGTH FULL SIZE TABLE 1305.2
BRANCH VENT
YOKE VENT FOR OFFSET RELIEF. 1417.3
INDIVIDUAL VENTS. 1409.1
FIXTURES BACK TO BACK 1414.3.2
FIXTURE BRANCH (FOR MAXIMUM FIXTURE UNITS & SIZE, SEE TABLE 1305.1)

Figure 12.26 Venting system diagram, Figure 2 of the code. (*Reproduced from the 1991 edition of* The Standard Plumbing Code® *with permission of the copyright holder, Southern Building Code Congress, International, Inc. All rights reserved.*)

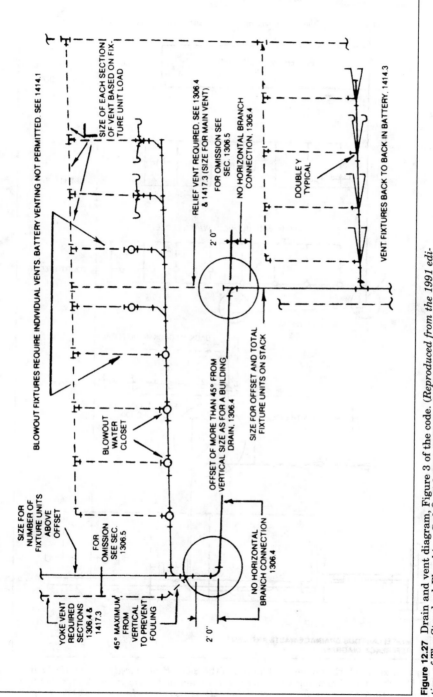

Figure 12.27 Drain and vent diagram, Figure 3 of the code. *(Reproduced from the 1991 edition of The Standard Plumbing Code® with permission of the copyright holder, Southern Building Code Congress, International, Inc. All rights reserved.)*

235

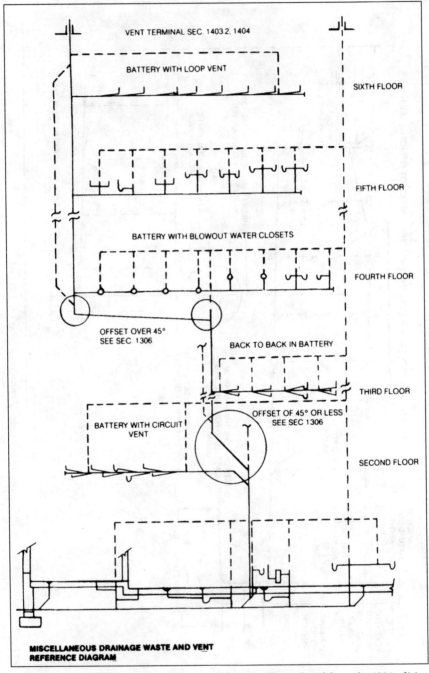

Figure 12.28 DWV riser diagram, Figure 1 of the code. (*Reproduced from the 1991 edition of* The Standard Plumbing Code® *with permission of the copyright holder, Southern Building Code Congress, International, Inc. All rights reserved.*)

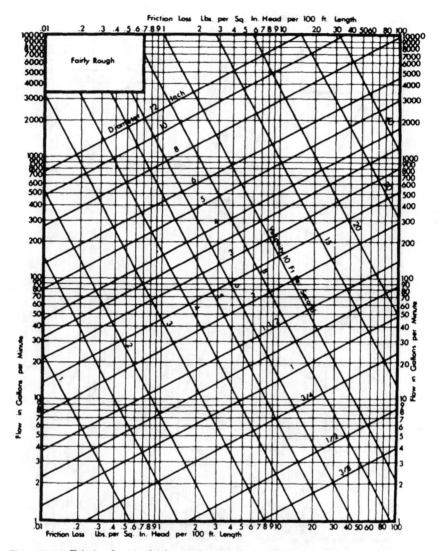

Figure 12.29 Friction loss in fairly rough pipe, Figure F103C of the code. (*Reproduced from the 1991 edition of* The Standard Plumbing Code® *with permission of the copyright holder, Southern Building Code Congress, International, Inc. All rights reserved.*)

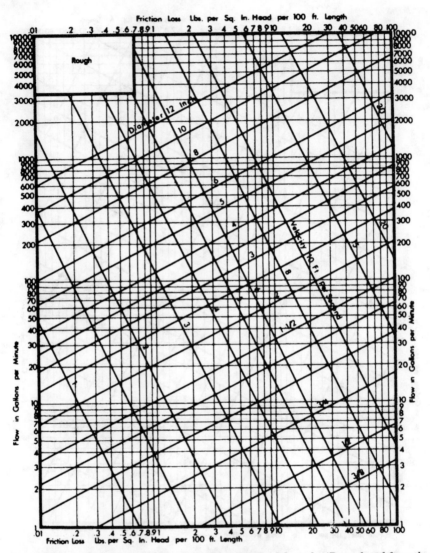

Figure 12.30 Friction loss in rough pipe, Figure 5103D of the code. (*Reproduced from the 1991 edition of* The Standard Plumbing Code® *with permission of the copyright holder, Southern Building Code Congress, International, Inc. All rights reserved.*)

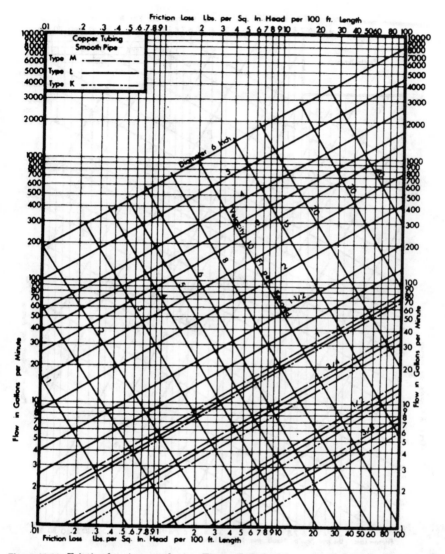

Figure 12.31 Friction loss in smooth pipe, Figure F103A of the code. (*Reproduced from the 1991 edition of* The Standard Plumbing Code® *with permission of the copyright holder, Southern Building Code Congress, International, Inc. All rights reserved.*)

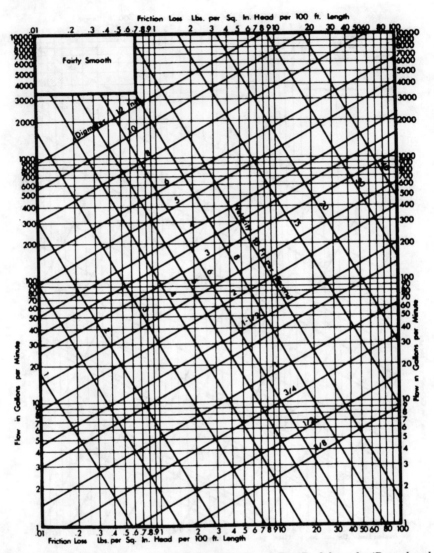

Figure 12.32 Friction loss in fairly smooth pipe, Figure F103B of the code. (*Reproduced from the 1991 edition of* The Standard Plumbing Code® *with permission of the copyright holder, Southern Building Code Congress, International, Inc. All rights reserved.*)

Gallons per	Size of tap or tee (in)						
minute	⅝	¾	1	1¼	1½	2	3
10	1.35	0.64	0.18	0.08			
20	5.38	2.54	0.77	0.31	0.14		
30	12.1	5.72	1.62	0.69	0.33	0.10	
40		10.2	3.07	1.23	0.58	0.18	
50		15.9	4.49	1.92	0.91	0.28	
60			6.46	2.76	1.31	0.40	
70			8.79	3.76	1.78	0.55	0.10
80			11.5	4.90	2.32	0.72	0.13
90			14.5	6.21	2.94	0.91	0.16
100			17.94	7.67	3.63	1.12	0.21
120			25.8	11.0	5.23	1.61	0.30
140			35.2	15.0	7.12	2.20	0.41
150				17.2	8.16	2.52	0.47
160				19.6	9.30	2.92	0.54
180				24.8	11.8	3.62	0.68
200				30.7	14.5	4.48	0.84
225				38.8	18.4	5.67	1.06
250				47.9	22.7	7.00	1.31
275					27.4	7.70	1.59
300					32.6	10.1	1.88

Figure 12.33 Loss of pressure through taps and tees, Table F103A of the code. (*Reproduced from the 1991 edition of* The Standard Plumbing Code® *with permission of the copyright holder, Southern Building Code Congress, International, Inc. All rights reserved.*)

	Pipe sizes (in)							
Fitting or valve	½	¾	1	1¼	1½	2	2½	3
45° elbow	1.2	1.5	1.8	2.4	3.0	4.0	5.0	6.0
90° elbow	2.0	2.5	3.0	4.0	5.0	7.0	8.0	10.0
Tee, run	0.6	0.8	0.9	1.2	1.5	2.0	2.5	3.0
Tee, branch	3.0	4.0	5.0	6.0	7.0	10.0	12.0	15.0
Gate valve	0.4	0.5	0.6	0.8	1.0	1.3	1.6	2.0
Balancing valve	0.8	1.1	1.5	1.9	2.2	3.0	3.7	4.5
Plug-type cock	0.8	1.1	1.5	1.9	2.2	3.0	3.7	4.5
Check valve, swing	5.6	8.4	11.2	14.0	16.8	22.4	28.0	33.6
Globe valve	15.0	20.0	25.0	35.0	45.0	55.0	65.0	80.0
Angle valve	8.0	12.0	15.0	18.0	22.0	28.0	34.0	40.0

Figure 12.34 Allowance in equivalent length of pipe for friction loss in valves and threaded fittings, Table F103B of the code. (*Reproduced from the 1991 edition of* The Standard Plumbing Code® *with permission of the copyright holder, Southern Building Code Congress, International, Inc. All rights reserved.*)

Fitting or valve	Tube sizes (in)							
	½	¾	1	1¼	1½	2	2½	3
45° elbow (wrought)	0.5	0.5	1.0	1.0	2.0	2.0	3.0	4.0
90° elbow (wrought)	0.5	1.0	1.0	2.0	2.0	2.0	2.0	3.0
Tee, run (wrought)	0.5	0.5	0.5	0.5	1.0	1.0	2.0	
Tee, branch (wrought)	1.0	2.0	3.0	4.0	5.0	7.0	9.0	
45° elbow (cast)	0.5	1.0	2.0	2.0	3.0	5.0	8.0	11.0
90° elbow (cast)	1.0	2.0	4.0	5.0	8.0	11.0	14.0	18.0
Tee, run (cast)	0.5	0.5	0.5	1.0	1.0	2.0	2.0	2.0
Tee, branch (cast)	2.0	3.0	5.0	7.0	9.0	12.0	16.0	20.0
Compression Stop	13.0	21.0	30.0					
Globe valve	7.5	10.0	12.5	53.0	66.0	90.0	33.0	40.0
Gate valve	0.5	0.25	1.0	1.0	2.0	2.0	2.0	2.0

Figure 12.35 Allowance in equivalent length of tube for friction loss in valves and fittings, Table F103C of the code. (*Reproduced from the 1991 edition of* The Standard Plumbing Code® *with permission of the copyright holder, Southern Building Code Congress, International, Inc. All rights reserved.*)

Nominal internal diameter of pipe (in)	Discharge units
2	20 (No W.C.)
2½	80 (No W.C.)
3	200 (No W.C.)
4	850
5	2700
6	6500

[1] The capacity of a vertical discharge pipe (stack) is limited by the need to preserve a large air core to prevent excessive pressure fluctuation. The flow capacity of a stack may therefore be less than that of a pipe of the same diameter laid at a steep fall.

[2] Discharge pipes sized by this method give the minimum size necessary to carry the expected flow load. Separate ventilation pipes may be required (see 1602.6). It may be worthwhile to consider oversizing the discharge pipes to reduce the ventilating pipework required.

Figure 12.36 Maximum number of discharge units allowed on vertical stacks, Table 1602.4B of the code.[1,2] (*Reproduced from the 1991 edition of* The Standard Plumbing Code® *with permission of the copyright holder, Southern Building Code Congress, International, Inc. All rights reserved.*)

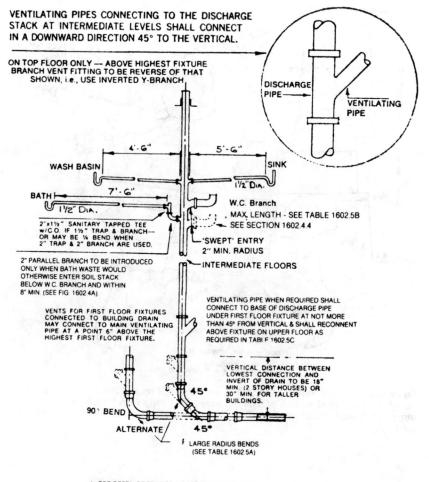

Figure 12.37 Main features of single stack system, Figure 1602.4B of the code. (*Reproduced from the 1991 edition of* The Standard Plumbing Code® *with permission of the copyright holder, Southern Building Code Congress, International, Inc. All rights reserved.*)

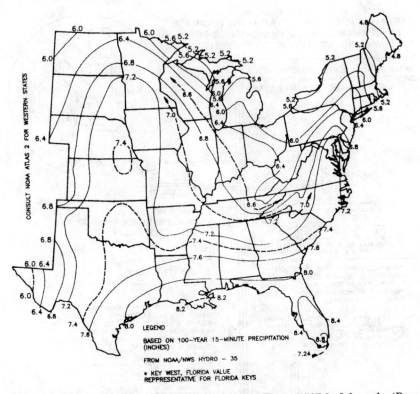

Figure 12.38 Rainfall rates for secondary roof drains, Figure 1507.3 of the code. (*Reproduced from the 1991 edition of* The Standard Plumbing Code® *with permission of the copyright holder, Southern Building Code Congress, International, Inc. All rights reserved.*)

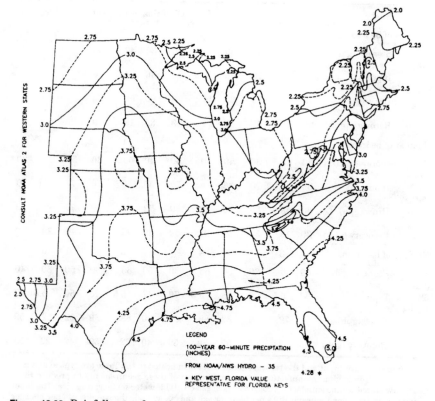

Figure 12.39 Rainfall rates for primary roof drains, Figure 1506.4 of the code. (*Reproduced from the 1991 edition of* The Standard Plumbing Code® *with permission of the copyright holder, Southern Building Code Congress, International, Inc. All rights reserved.*)

Fuel	Gas	Elec.	Oil	Gas	Elec.	Oil	Gas	Elec.	Oil	Gas	Elec.	Oil
Number of bedrooms		1			2			3			—	
1 to 1½ Baths Storage (gal)	20	20	30	30	30	30	30	40	30			
Input	27	2.5	70	36	3.5	70	36	4.5	70			
Draw (gph)	43	30	89	60	44	89	60	58	89			
Recovery (gph)	23	10	59	30	14	59	30	18	59			
Number of bedrooms		2			3			4			5	
2 to 2½ Baths Storage (gal)	30	40	30	40	50	30	40	50	30	50	66	30
Input	36	4.5	70	36	5.5	70	38	5.5	70	47	5.5	70
Draw (gph)	60	58	89	70	72	89	72	72	89	90	88	89
Recovery (gph)	30	18	59	30	22	59	32	22	59	40	22	59
Number of bedrooms		3			4			5			6	
3 to 3½ Baths Storage (gal)	40	50	30	50	66	30	50	66	30	50	80	40
Input	38	5.5	70	38	5.5	70	47	5.5	70	50	5.5	70
Draw (gph)	72	72	89	82	88	89	90	88	89	92	102	99
Recovery (gph)	32	22	59	32	22	59	40	22	59	42	22	59

[1] Storage capacity, input and the recovery requirements indicated in the Table are typical and may vary with each individual manufacturer. Any combination of these requirements to produce the 1 hour draw stated will be satisfactory. Recovery is based on 100 F water temperature rise. The input rating is in units of one thousand Btu's per hour for gas and oil, and one thousand watts per hour for electric.

Example: For a 3-bedroom, 2-bath residence there are three choices as follows: A 40 gal storage/30 gph recovery gas heater; a 50 gal storage/22 gph recovery electric heater; or a 30 gal storage/59 gph recovery oil heater; or an equivalent combination which will produce at least a 70 gph total draw.

Figure 12.40 Minimum water heater capacities, Table 921.1.5 of the code.[1] (*Reproduced from the 1991 edition of* The Standard Plumbing Code® *with permission of the copyright holder, Southern Building Code Congress, International, Inc. All rights reserved.*)

Building or occupancy[2]	Occupant content[2]	Water closets[3]			Lavatories[4]			Bathtubs, showers, and miscellaneous fixtures
Dwelling or apt. house	Not applicable	1 for each dwelling or dwelling unit			1 for each dwelling or dwelling unit			Washing machine connection per unit.[5] Bathtub or shower—one per dwelling or dwelling unit. Kitchen sink—One per dwelling or dwelling unit
Schools: preschool, day care, or nursery	Average daily attendance	Each 15 children or fraction thereof · 1 Fixture			Each 15 children or fraction thereof 1 Fixture			
Schools: Elementary & secondary	Average daily attendance	Persons (total)	Male	Female	Persons (total)	Male	Female	One drinking fountain for each 3 classrooms, but not less than one each floor.
		1–50	2	2	1–120	1	1	
		51–100	3	3	121–240	2	2	
		101–150	4	4	For each additional 120 persons over 240, add	1	1	
		151–200	5	5				
		For each additional 50 persons over 200, add	1	1				

		Water closets[3]			Lavatories[4]			Drinking Fountains	
Office[6] and public buildings	100 sq ft per person	Persons (total)	Male	Female	Persons (total)	Male	Female	Persons	Fixtures
		1–15	1	1	1–15	1	1	1–100	1
		16–35	1	2	16–35	1	2	101–250	2
		36–55	2	2	36–60	2	2	251–500	3
		56–100	2	3	61–125	2	3	Not less than one fixture each floor subject to access.	
		101–150	3	4	For each additional 120 persons over 125, add	1	1.5[7]		
		For each additional 100 persons over 150, add	1	1.5[7]					

Figure 12.41 Minimum plumbing fixtures, Table 922.2 of the code. (Reproduced from the 1991 edition of The Standard Plumbing Code® with permission of the copyright holder, Southern Building Code Congress, International, Inc. All rights reserved.)

Building or occupancy[2]	Occupant content[2]	Water closets[3] Persons (total)	Male	Female	Lavatories[4] Persons (total)	Male	Female	Bathtubs, showers, and miscellaneous fixtures — Drinking Fountains Persons	Fixtures
Common toilet facilities or areas of commercial buildings of multiple tenants[8,9]	Use the sq ft per person ratio applicable to the single type occupancy(s) occupying the greatest aggregate floor area (Consider separately each floor area of a divided floor)	1–50 / 51–100 / 101–150	2 / 3 / 4	2 / 4 / 5	1–15 / 16–35 / 36–60 / 61–125	1 / 1 / 2 / 2	1 / 2 / 2 / 3	1–100 / 101–250 / 251–500 / 501–1000	1 / 2 / 3 / 4
		For each additional 100 persons over 150, add	1	1.5[7]	For each additional 120 persons over 125, add	1	1.5[7]	Not less than one fixture each floor subject to access.	
Retail stores[6]	200 sq ft per person	1–35 / 36–55 / 56–80 / 81–100 / 101–150	1 / 1 / 2 / 2 / 2	1 / 2 / 3 / 4 / 5	1–15 / 16–35 / 36–60 / 61–125	1 / 1 / 1 / 2	1 / 2 / 3 / 4	1–100 / 101–250 / 251–500 / 501–1000	1 / 2 / 3 / 4
		For each additional 200 persons over 150, add	1	1.75[7]	For each additional 200 persons over 125, add	1	1.75[7]	Not less than one fixture each floor subject to access.	
Restaurants,[6] clubs, and lounges	40 sq ft per person	1–50 / 51–100 / 101–300	2 / 3 / 4	2 / 3 / 4	1–150 / 151–200 / 201–400	1 / 2 / 3	1 / 2 / 3	Comply with Board of Health Requirements.	
		For each additional 200 persons over 300, add	1	2	For each additional 200 persons over 400, add	1	1		

Figure 12.42 Minimum plumbing fixtures, Table 922.2, continued, of the code. (Reproduced from the 1991 edition of The Standard Plumbing Code® with permission of the copyright holder, Southern Building Code Congress, International, Inc. All rights reserved.)

Figure 12.43 Minimum plumbing fixtures, Table 922.2, continued, of the code. (*Reproduced from the 1991 edition of The Standard Plumbing Code® with permission of the copyright holder; Southern Building Code Congress, International, Inc. All rights reserved.*)

Building or occupancy[2]	Occupant content[2]	Water closets[3]			Lavatories[4]			Bathtubs, showers, and miscellaneous fixtures
		Persons (total)	Male	Female	Persons (total)	Male	Female	
Do it yourself laundries[6]	50 sq ft per person	1–50 51–100	1 1	1 2	1–100	1	1	One drinking fountain and one service sink.
Beauty shops and barber shops[6]	50 sq ft per person	1–35 36–75	1 2	1 2	1–75	1	1	One drinking fountain and one service or other utility sink.
Heavy manufacturing,[10] warehouses,[11] foundries, and similar establishments[12,14]	Occupant content per shift, substantiated by owner. Also see 922.3.2	1–10 11–25 26–50 51–75 76–100 For each additional 60 persons over 100, add	1 2 3 4 5 1	1 1 1 1 1 0.1[7]	1–15 16–35 36–60 61–90 91–125 For each additional 100 persons over 125, add	Male[14] 1 2 3 4 5 1	Female[14] 1 1 1 1 1 0.1[7]	One drinking fountain for each 75 persons. One shower for each 15 persons exposed to excessive heat or to skin contamination with poisonous, infectious, or irritating material.
Light mfg.,[10] Light warehousing,[11] and workshops, etc.[12,13]	Occupant content per shift, substantiated by owner. Also see 922.3.2	1–25 26–75 76–100 For each additional 60 persons over 100, add	1 2 3 1	1 2 3 1	1–35 36–100 101–200 For each additional 100 persons over 200, add	Male[14] 1 2 3 1	Female[14] 1 2 3 1	One drinking fountain for each 75 persons. One shower for each 15 persons exposed to excessive heat or to skin contamination with poisonous, infectious, or irritating material.

249

Building or occupancy[2]	Occupant content[2]	Water closets[3]			Lavatories[4]			Bathtubs, showers, and miscellaneous fixtures
		Persons (total)	Male[16]	Female[16]	Persons (total)	Male[16]	Female[16]	
Dormitories[15]	50 sq ft per person (calculated on sleeping area only)	1–10	1	1	1–12	1	1	Washing machines may be used in lieu of laundry tubs.[15] One shower for each 8 persons. In women's dorms add tubs in the ratio 1 for each 30 females. Over 150 persons add 1 shower for each 20 persons.
		11–30	1	2	13–30	2	2	
		31–100	3	4	For each additional 30 persons over 30, add	1	1	
		For each additional 50 persons over 100, add	1	1				
Theaters, auditoriums, churches, waiting rooms at transportation terminals, and stations	70 sq ft per person (calculated from assembly area). Other areas considered public buildings.	1–50	2	2	1–200	1	1	
		51–100	3	3	201–400	2	2	
		101–200	4	4	401–750	3	3	
		201–400	5	5	For each additional 350 persons over 750, add	1	1	
		For each additional 250 persons over 400, add	1	1				

Drinking Fountains

Persons	Fixtures
1–100	1
101–350	2
Over 350 add one fixture for each 400.	

Figure 12.44 Minimum plumbing fixtures, Table 922.2, continued, of the code. (*Reproduced from the 1991 edition of The Standard Plumbing Code® with permission of the copyright holder; Southern Building Code Congress, International, Inc. All rights reserved.*)

[1] The figures shown are based upon one fixture being the minimum required for the number of persons indicated or any fraction thereof.

[2] The occupant content and the number of required facilities for occupancies other than listed shall be determined by the Plumbing Official. Plumbing facilities in the occupances or tenancies of similar use may be determined by the Plumbing Official from this table.

[3] Urinals shall be required in male restrooms of elementary or secondary schools, restaurants, clubs, lounges, waiting rooms of transportation terminals, auditoriums, theaters, and churches at a rate equal to ⅓ of the required water closets in Table 922.2. Required urinals can be substituted for up to ⅓ of the required water closets. The installation of urinals shall be optional in the female restrooms of previously stated occupancies and shall be optional in both male and female restrooms of all other occupancies. Optional urinals may be substituted for up to ½ of the required water closets in the male and female restrooms.

[4] Twenty-four linear inches of wash sink or 18 inches of a circular basin, when provided with water outlets for such space, shall be considered equivalent to 1 lavatory.

[5] When central washing facilities are provided in lieu of washing machine connections in each living unit, central facilities shall be located for the building served at the ratio of not less than one washing machine for each 12 living units, but in no case less than two machines for each building of 15 living units or less. See 914.5.

[6] A single facility consisting of one water closet and one lavatory may be used by both males and females in the following occupancies subject to the building area limitations:

Occupancy	Maximum Building Area (sq ft)
Office	1200
Retail Store (excluding service stations)	1500
Restaurant	500
Laundries (Self Service)	1400
Beauty and Barber Shops	900

Figure 12.45 Footnotes to Table 922.2 of the code. (*Reproduced from the 1991 edition of The Standard Plumbing Code® with permission of the copyright holder, Southern Building Code Congress, International, Inc. All rights reserved.*)

Fixture type	Fixture-unit value as load factors	Minimum size of trap (in)
Bathroom group consisting of water closet, lavatory, and bathtub or shower	6	
Bathtub[1] (with or without overhead shower) or whirlpool attachments	2	1½
Bidet	2	Nominal 1½
Combination sink and tray	3	1½
Combination sink and tray with food disposal unit	4	Separate traps 1½
Dental unit or cuspidor	1	1¼
Dental lavatory	1	1¼
Drinking fountain	½	1
Dishwashing machine[2] domestic.	2	1½
Floor drains[5]	1	2
Kitchen sink, domestic	2	1½
Kitchen sink, domestic with food waste grinder and/or dishwasher	3	1½
Lavatory[4]	1	Small P.O. 1¼
Lavatory[4]	2	Large P.O. 1½
Lavatory, barber, beauty parlor	2	1½
Lavatory, surgeon's	2	1½
Laundry tray (1 or 2 compartments)	2	1½
Shower stall, domestic	2	2
Showers (group) per head[2]	3	
Sinks		
Surgeon's	3	1½
Flushing rim (with valve)	8	3
Service (trap standard)	3	3
Service (P trap)	2	2
Pot, scullery, etc.[2]	4	1½
Urinal, pedestal, siphon jet, blowout	8	Note 6
Urinal, wall lip	4	Note 6
Urinal, washout	4	Note 6
Washing machines (commercial)[3]		
Washing machine (residential)	3	2
Wash sink[2] (circular or multiple) each set of faucets	2	Nominal 1½
Water closet, flushometer tank, public or private	3	Note 6
Water closet, private installation	4	Note 6
Water closet, public installation	6	Note 6

[1] A showerhead over a bathtub or whirlpool bathtub attachments does not increase the fixture value.

[2] See 1304.2 and 1304.3 for methods of computing unit value of fixtures not listed in Table 1304.1 or for rating of devices with intermittent flows.

[3] See Table 1304.2.

[4] Lavatories with 1¼ or 1½-inch trap have the same load value; larger P.O. plugs have greater flow rate.

[5] Size of floor drain shall be determined by the area of the floor to be drained. The drainage fixture unit value need not be greater than 1 unless the drain receives indirect discharge from plumbing fixtures, air conditioner, or refrigeration equipment.

[6] Trap size shall be consistent with fixture type as defined in industry standards.

Figure 12.46 Fixture units per fixture or group, Table 1304.1 of the code. (*Reproduced from the 1991 edition of* The Standard Plumbing Code® *with permission of the copyright holder, Southern Building Code Congress, International, Inc. All rights reserved.*)

Fixture drain or trap size (in)	Fixture-unit value
1¼ and smaller	1
1½	2
2	3
2½	4
3	5
4	6

Figure 12.47 Fixtures not listed, Table 1304.2 of the code. *(Reproduced from the 1991 edition of* The Standard Plumbing Code® *with permission of the copyright holder, Southern Building Code Congress, International, Inc. All rights reserved.)*

		Maximum no. of fixture units that may be connected to:		
		One stack of 3 stories	More than 3 stories in height	
Diameter of pipe[5] (in)	Any horizontal fixture branch[1,4]	or 3 intervals maximum	Total for stack	Total at one story or branch interval
1¼	1	2	2	1
1½	3	4	8	2
2	6	10	24	6
2½	12	20	42	9
3	20[2]	30[3]	60[3]	16[2]
4	160	240	500	90
5	360	540	1100	200
6	620	960	1900	350
8	1400	2200	3600	600
10	2500	3800	5600	1000
12	3900	6000	8400	1500
15	7000			

[1] Does not include branches of the building drain.
[2] Not over two water closets.
[3] Not over six water closets.
[4] 50% less for battery vented fixture branches, no size reduction permitted for battery vented branches throughout the entire branch length.
[5] The minimum size of any branch or stack serving a water closet shall be 3″.

Figure 12.48 Maximum number of fixture units that may be connected to horizontal fixture branches and stacks, Table 1305.2 of the code. *(Reproduced from the 1991 edition of* The Standard Plumbing Code® *with permission of the copyright holder, Southern Building Code Congress, International, Inc. All rights reserved.)*

Diameter of pipe (in)	Fall in inches per foot			
	$\frac{1}{16}$	$\frac{1}{8}$	$\frac{1}{4}$	$\frac{1}{2}$
2			21	26
2½			24	31
3		20²	27²	36²
4		180	216	250
5		390	480	575
6		700	840	1000
8	1400	1600	1920	2300
10	2500	2900	3500	4200
12	3900	4600	5600	6700
15	7000	8300	10,000	12,000

[1] Includes branches of the building drain. The minimum size of any building drain serving a water closet shall be 3″.
[2] Not over two water closets.

Figure 12.49 Building drains and sewers, maximum number of fixture units that may be connected to any portion of the building drain or building sewer, Table 1305.1 of the code.[1] (*Reproduced from the 1991 edition of* The Standard Plumbing Code® *with permission of the copyright holder, Southern Building Code Congress, International, Inc. All rights reserved.*)

Size of fixture drain (in)	Size of trap (in)	Fall (in/ft)	Max. distance from trap
1¼	1¼	¼	3 ft 6 in
1½	1¼	¼	5 ft
1½	1½	¼	5 ft
2	1½	¼	8 ft
2	2	¼	6 ft
3	3	⅛	10 ft
4	4	⅛	12 ft

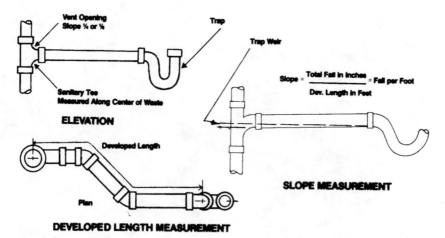

Figure 12.50 Distance of fixture trap from vent, Table 1408.1 of the code. (*Reproduced from the 1991 edition of* The Standard Plumbing Code® *with permission of the copyright holder, Southern Building Code Congress, International, Inc. All rights reserved.*)

Number of wet-vented fixtures	Diameter of vent stacks (in)
1 or 2 bathtubs or showers	2
3 to 5 bathtubs or showers	2½
6 to 9 bathtubs or showers	3
10 to 16 bathtubs or showers	4

Figure 12.51 Size of vent stack, Table 1411.3 of the code. (*Reproduced from the 1991 edition of* The Standard Plumbing Code® *with permission of the copyright holder, Southern Building Code Congress, International, Inc. All rights reserved.*)

13

Preparing for the Plumber's Licensing Exam

The most effective way to pass the plumber's licensing exam is to be properly prepared. There is more to proper preparation than just studying the code book and remembering what you have studied. Mental attitude has a great deal to do with an individual's performance. Mind-blocking can be necessary for experienced plumbers sitting for their test. These are only a few of the factors that will influence your exam scores.

This chapter is going to share with you my experiences as a plumber, plumbing contractor, and code instructor. In sharing my experience, you will not only learn from my mistakes but from mistakes made by the many people I have guided to successful completions in licensing exams. If you plan to take a test for a plumber's license, this chapter will help you pass the test.

Mind-Blocking

Mind-blocking is often one of the first requirements for experienced plumbers preparing to sit for a licensing exam. It is the act of blocking out preconceived notions and experience so your mind can accept the proper learning procedures. It is often more difficult for experienced plumbers to pass a licensing exam than it is for rookies because experienced plumbers think they know how to do plumbing. It is a fact that these experienced plumbers do know how to get the job done, but that doesn't mean they know how to perform on paper. There is a big difference between plumbing in the field and plumbing on paper.

We all know that life does not follow a text book. In the real world, things are often done in a manner that contradicts the written word. All experienced plumbers have learned to make on-site adjustments. They know what their local code officers are looking for, and their jobs pass inspection because of this knowledge. However, what the local code officer is looking for is not always what the licensing exam is looking for.

When taking the licensing exam, you must put everything out of your mind except the proper procedures for passing the exam. You cannot assume you know the code simply because your jobs always pass inspection. How often does the average plumber size roof drains? Not very often, but on the licensing exam, roof drains are likely to be there. A majority of plumbers never plumb slaughterhouses or morgues, but the test may require an exhibition of knowledge in these areas.

Since many plumbers specialize, it is not uncommon for a residential plumber to have a very limited knowledge of commercial plumbing procedures. The requirements between residential and commercial plumbing can vary a great deal; even if you never plan to do commercial work, you will have to learn the basics to pass most licensing exams.

The biggest hurdle to overcome is the one of bad habits. Experienced plumbers tend to fall into a pattern. Once they know what will be accepted by their local code officials, they stick with the procedure. This is fine for day-to-day working, but it can result in failure of the licensing exam. Let me give you a few examples of how these comfortable patterns can cause you to fail your test.

In day-to-day field work, it is common to secure pipe hangers to floor joists with galvanized roofing nails. Galvanized nails are used frequently with copper hangers and clips. But, if you check code requirements, you will find that hangers must be attached with a device, like a nail, that will not cause a corrosive reaction with the hanger. Technically, copper nails should be used with copper hangers, but how many working plumbers do this? Not many, but they are allowed to continue using galvanized nails because inspectors don't reject the jobs.

If these plumbers are given a multiple-choice question on nail selection, what type of nail to you think they will pick? They will generally pick the type of nail they are accustomed to using. In the real world, their selection passes the average inspection, but on the licensing exam, choosing galvanized nails for use with copper hangers will result in an incorrect answer.

Let me give you another example. I have worked in many jurisdictions where inspectors were lenient on the testing of new drain waste and vent (DWV) systems. There have been many times when I was allowed to test my pipes by filling up the bathtub on the highest level of the building. The premise for this was that if a drain was going to leak, it would be found by filling the highest drainage fixture with water. But, what about the vent pipes that are above the highest fixture. Really, all of these pipes should be tested.

Most codes require all DWV pipes to be tested from their lowest point to their highest point, the highest point generally being the vent terminal on the roof. Now for years I was allowed to test by merely filling the highest bathtub. It would be easy to assume this was all the code required; after all, my jobs passed inspections. However, if a test question was answered on this assumption, it would be wrong.

Experienced plumbers must be willing to take the test on its own terms. Regardless of what works in the field, only the correct answers work on the exam. The problem of putting aside daily habits to answer test questions with the answers wanted has been a major problem for many experienced plumbers taking the licensing exam.

Mental Attitude

Mental attitude may be the single most important element in preparing for a licensing exam. If you are too nervous, you are unlikely to pass the test. If you too self-assured, you are likely to be surprised at the questions you are faced with. Ideally, you want to strive for the thin line between being afraid and being cocky. The key word is respect. You must respect the licensing exam. It can be formidable and intimidating if you are not well prepared.

When preparing for the test, review your study material, but don't dwell on it to the point where it begins to blur. When you are taking the test, avoid second-guessing your answers. A lot of exam applicants finish their test early and go back through it, changing answers they are unsure of. Unless you are sure your first answer is wrong, and I mean sure, don't change it. The odds are good you answered it correctly the first time. If you have prepared properly for the test, arriving at the correct answers will be almost an instinct.

It is wise to check over your math computations to make sure no mistakes were made in the mechanical application of your math. But otherwise, if you have learned your material, the right answers will jump out of your brain and onto the paper. Going back and changing initial answers will normally create more incorrect answers than correct ones.

Calculators

If your testing agency allows the use of an electronic calculator, use one that you are familiar with. Don't buy a new calculator on the day of the exam and try to learn to use it while taking the exam. The exam will be taxing enough without the added distraction of an unfamiliar calculator.

Code-Preparation Courses

As I have mentioned earlier in the book, code-preparation courses can be the cornerstone of a successful exam completion. Many plumbers sitting for their licensing exams have been out of school for a long time. When you are away from the demands of studying and testing, you forget how to do it. Code courses help you redevelop good study habits. You are required to study and take in-class tests. This experience is very valuable when it is time to take the real exam.

Another benefit of code classes is the opportunity to work with others. If you sit and stare at a code book, you can become disinterested quickly. By studying in a group environment, with a decent instructor, learning is much less painful. This type of learning experience also reveals different interpretations of the plumbing code. Since much of the code is a matter of interpretation, it helps to have the opinions of others in determining what the code is trying to say.

The I've-Got-to-Pass-It Syndrome

Many exam applicants doom themselves to failure from the beginning. They put undue stress on themselves with what I call the I've-got-to-pass-it syndrome. Passing a licensing exam is not difficult when you are properly prepared. However, if you come to the testing table with your mind concentrating on the negative effects of failing the test, you will probably fail.

Some people's nature will not allow them to avoid the I've-got-to-pass-it syndrome. For these people, the best alternative is to take a close look at the true ramifications of failure. In reality, many plumbers pass these exams on their first attempt. Generally, the exams are given at regular intervals, only a month or so apart. What is the worst that can happen if you fail the test? You must take it again. But, you will be better prepared the next time. The test will not be the same, but it will be similar. You have seen it once; you are better equipped to study for it the second time around.

In all probability, you have waited at least 2 years to take your first test. How much difference will it make if you have to wait a month to take another test? Put the importance of the exam into perspective. Try to avoid becoming daunted by the fear of failure. If you can clear your mind of negative thoughts, the lessons you have studied will shine through.

The Bottom Line

The bottom line is to be well prepared for your licensing exam. If you need to take a day off from work to get mentally prepared, do it. If you have trouble studying on your own, join a study group or code class. Don't try to guess what will be on the test; learn the code, and you will not have to worry about what is on the exam. Don't listen to the remarks of others who have recently taken the exam. Licensing exams are changed regularly. What your coworkers encountered on their tests may not be on yours.

The keys to success are knowledge and preparation. Except for a rare few, who freeze on tests, if you know the code, you will pass the test. I have taken many licensing exams. I have taking them for plumbing and real estate licenses. I have taken them in different states. The last master plumber's test I took, I only had 2 days to study for. This test was administered when I moved to Maine, and I had no idea I would get the opportunity to take it on such short notice.

I have never failed a licensing exam. My scores have always been good, and I've always passed on my first attempt. What's my secret? I take the time to learn the required material, and I condition myself to have a positive attitude when taking the exams. Have I ever been afraid of failing my exams? Only every time I take them, but I overcome my fear with knowledge. If you know the material, you can pass the test. In most cases, you can forget 25 percent of what you need to know and still pass.

Licensing exams are not the monsters they are made out to be. They are blown out of proportion by people who didn't spend the effort to study for them. The next chapter is going to share some study techniques with you. I am confident you can pass your exam, if you respect the test and prepare for it accordingly.

14

Effective Study Methods

Effective study methods are instrumental in the successful completion of a plumber's licensing exam. Opening a code book and staring at it will not get the job done. In most cases, memorizing the code text is not enough to pass a licensing exam. Licensing exams are intended to test an individual's ability to work with the plumbing code. There is a big difference between committing the code to memory and being able to apply the code to a licensing exam.

Some exam questions can be answered satisfactorily by remembering the code, word for word, but many of the questions will require individuals to apply their knowledge of the code in determining a correct answer. Pipe sizing is a good example. A person can memorize the elements needed to size pipe, but if that person cannot use the elements to perform the math, the answer to the test question is likely to be wrong. The same circumstances could apply to other questions that rely on the successful completion of formulas and math to obtain a correct answer.

The plumbing code is a complex compilation of information. Trying to pass a licensing exam without a thorough knowledge of the code is likely to end in failure. This is not to say you must know every aspect of the code, inside and out. But, a successful license applicant must know the basics and how to apply them.

The last chapter dealt with preparing to take a licensing exam. This chapter expands on that preparation, with information on effective study methods. Learning techniques that work for one person may not work for another. The information that follows in this chapter is based on my experience as a plumbing instructor and as a plumber and employer. Through my years of experience I have helped numerous people obtain their plumber's license. I have also seen many fail. What follows are suggestions from my personal experience and from experiences shared with me by others over the years.

Code Classes

We have discussed code classes earlier, so I won't spend a lot of time on them here, but I do consider them a helpful element in passing licensing exams. Code classes are designed to teach you the plumbing code. In some cases, they are aimed at preparing you to pass a licensing exam. These classes offer the benefit of group study, mock exams, and experienced instructors.

If you have trouble disciplining yourself to study, code courses can help. If you are unable to study at home, because of common distractions, code courses are great. When you are not sure how to interpret the code, it is very beneficial to have an experienced instructor to question. In general, code courses are a good value and provide much assistance in learning the plumbing code.

Studying with a Friend

Studying with a friend allows for better self-testing of your knowledge. If you don't attend a code class, studying with a friend is the next best approach. It helps if the friend is knowledgeable of the plumbing code, but it is not mandatory. Almost anyone is capable of assisting in your learning goals.

There are two common approaches to studying with a friend. Your assistant can quiz you verbally or with written tests. Most study teams use verbal testing, but it is best to use a combination of both methods. Let me give you some examples of how you might structure your study time with a friend.

Take the plumbing code glossary as an example. To be prepared for the licensing exam, you should know all of the glossary words and their meanings. It is too easy to inadvertently cheat when testing yourself on this type of information. Have your study partner go through the glossary and ask you to define the words, at random. Occasionally have your assistant read a definition and ask you for the proper word. Mixing up the method of asking the questions will make memorization less likely and learning more likely.

Have your partner skip around the glossary. Don't go down the list alphabetically. When a routine is established, it is too easy to get caught in the memorization trap. By the day of the test, your memory may not respond to the methods in which the test questions are answered. By learning the words, instead of memorizing them, you are better prepared to pass your exam.

If you have a willing study friend, have the friend go through the glossary and prepare some written quizzes. You can use a true-false or a multiple-choice format. The written quizzes often do more for the learning process than oral tests.

Use this same type of procedure to work your way through the code book. Remember to have your study partner move about through the book, at random, to simulate a real test. For example, have three questions on drainage systems and then three questions on water distribution.

In the beginning, when you are first learning the code, going through the book in chronological order is fine. Your chances of learning initially are better when your thoughts are concentrated on the same aspects of the code. But, when you are preparing for your exam, skip around and give your knowledge a true test.

Benefiting from Your Employer

If you have a good job, you receive many benefits from your employer. If your employer happens to be a master plumber, you may be able to derive study benefits from your boss. Be careful when taking the advice of a working plumber; remember what I said in the last chapter about mind-blocking and bad habits. However, if you have questions about the code, a master plumber may be able to help you.

Code Officers

Code officers are another potential source of help when you have questions about the code. Code officers should be a good authority on the proper code procedures, but not all code officers are. It is not just experienced plumbers who get set in their ways. Advice from a helpful code officer could be wrong. Always cross-reference any advice you get with your code book. Remember, you will be tested on text-book methods, not real-world procedures.

Self-Study

Self-study is probably the most popular way to study for a licensing exam. Self-study is good, but it does carry some potential pitfalls. As I mentioned when we were talking about studying the code glossary, it is easy to inadvertently cheat when you are forced to study alone.

If you are going to tutor yourself, play it straight. Cheating on your studies will only hurt you. I have found a method of self-study that is particularly effective. This is a principal I have used many times, and it has always worked. Let me show you my way for self-testing.

My special method of self-testing requires some preliminary work, before you begin to study. Basically, you are going to draft your sample quizzes and tests before you understand what you are testing yourself on. This way, it is nearly impossible to cheat on yourself.

Making your own tests

Before you begin studying, get your code book, a pencil, and a supply of paper. Using the glossary as an example, go through the glossary, at random, not in order, and jot down the words. At times, write out the definitions, without using the key word. Be sure to skip around in your word selection.

Then, thumb through the book, at random, and pretend you are a teacher, looking for test questions. You might write down a question that asks how far a 1½-in trap can be from its vent. You continue in this method until you have developed a couple hundred questions. When you have sufficiently covered the code book, put your sample tests away. Then, when you feel you have mastered the code, give yourself the tests you created.

By working in this manner, you are far more likely to obtain a fair evaluation of your retained knowledge. This may seem simple, or it may even seem stupid, but it works.

Set aside designated study time

To study effectively, you must set aside designated study time. Most of us have hectic schedules, but we are able to make time for the things we want to do. You may not want to study, but if you want to pass your licensing test, you had better make time to study. The length of time you arrange for is not as important as the intervals of study.

If you can only study for 30 min each night, fine, but study for those 30 min every night. It is generally not enough to say you will study at least 30 min each night. Most people will not stick to their commitment unless the study time is scheduled. Pretend you are going to a night class. Make arrangements to study every night at a specific time.

If you are supposed to study from 8:00 p.m. until 8:30 p.m., discipline yourself to do it. Ask your family members to cooperate in allowing you to be undisturbed during your study time. You will get far less done if you are interrupted during your study. If you simply cannot find peace and quiet at home, go to a library. If you want to learn badly enough, you will find the time and the means to do so.

Set goals

Set realistic goals for your study plan. For example, your goal might be to learn all the glossary words in your first three nights. It helps some people to create a goal chart. This can be little more than a piece of paper with grids and goals on it. The point is, make goals and strive to achieve them. If you get behind schedule, try to find a little more time to dedicate to your study. You might want to consider finding a study partner or code class if you are having significant trouble.

Repetition pays off

When studying or advertising, repetition pays off. The more we hear a slogan or read the code, the more we remember it. We are all aware of how repetition is used in advertising. Advertisers blitz us with their names, slogans, jingles, and logos. They do this for a good reason; they know if we see or hear their ads often enough, we will remember them. Even subconsciously, we remember facts we learn through repetition.

Do you remember the last time you used a combination lock or called your best friend? I'm willing to bet you were able to accomplish both of these tasks without referring to notes. But, when you first started using that combination or phone number, could you do it from memory? Probably not, but the more you use the number combinations, the easier they become to remember. The same is true of the plumbing code. The more you study and use the code, the easier it is to remember.

Don't take a quick pass through the code and assume you know it. When you feel you have learned the code, test yourself. Even if you do well on the tests, continue to review the code. The more you read and study the regulations, the better your odds for remembering them will become.

Study guides

Study guides can be a big help to you. Books that offer help in preparing for licensing exams give you a fresh angle of approach. Most of these books will have mock tests in them, like this one does. The chances are good that the tests in the books will test knowledge you have not tested. In the least, they should test the knowledge in a different way. This testing helps you in assessing the development of your understanding of the code.

Should you rely only on study guides? No, study guides are intended to help you learn the code, not to learn it for you. Use your code book and normal study methods to gain your base of code knowledge. When you think you know the code, use study guides to reinforce your feelings. After taking your own tests, try the tests in study guides. The more time you put into the learning process, the better off you will be. Not only will this type of study help you pass your exam, it will make you a better plumber.

Experiment

If you haven't studied for a long time and don't know where to begin, experiment with different methods. Some people find it easier to study when they are listening to a radio. Others watch television while they study. My preference is to study in a quiet, undisturbed area. However, this doesn't mean I study in a library. I enjoy studying outdoors. I find it relaxing to sit in the woods or in a meadow and study. I have used this method since high school, and it works for me. Experiment to find what conditions will allow you the most knowledge retention.

Allow Yourself Plenty of Time

It is important to allow yourself plenty of time for studying, before your test date. A lot of people put studying off until the last minute. This decision often results in failed tests. You should know months in advance when you will be eligible to sit for your exam. There is no valid excuse for waiting until the eleventh hour to begin studying.

One excuse I hear often is that if the person begins studying too soon, the information will be forgotten before the test date. Well, I don't buy this for two reasons. In the first place, you are supposed to be learning the code for more than just a test. You are intended to know the code to perform plumbing. If the information will be forgotten so soon, you will not make a very good plumber.

In the second place, the earlier you start, the more time you have for repetitive review. This repetitive review is the key to learning. My advice is to start early and review often. I believe that if you follow these suggestions, you will have little trouble in passing your exam.

15

Testing Your Knowledge

This chapter is going to test your knowledge. There are two types of tests in this chapter, true-false and multiple choice. Take these tests, and then look in the back of the chapter for the correct answers. The results of these tests will give you some insight as to how you would do on a licensing exam. These test questions are similar to the ones you might find on a licensing exam. In addition to these types of questions, most licensing exams will provide tables and charts for you to use in answering questions on pipe sizing.

This chapter gives you 50 questions and answers. Most licensing exams will have at least 100 questions. The questions and answers in this chapter are not meant to be representative of a licensing exam. They are meant to test your knowledge. Even if you ace these tests, you must still study for the licensing exam. If you don't do well on these tests, plan on doing extensive study before taking your licensing exam.

The answers that I have provided as being correct could vary with different code offices. Remember, all codes and code offices are not the same. However, the answers listed as being correct are accepted in most areas. Good luck on your tests.

Test One

Answer the following questions with true or false answers:

1. S-traps are legal for new installations so long as they are vented.

2. The minimum approved size for a water service is ½ in.

3. The minimum approved size for a water service is 1 in.

4. A full-open valve is required on the potable water inlet to a water heater.

5. A permit is not required when replacing an existing water heater.

6. Plans and specifications for work to be done may be required by the code enforcement office before a permit is issued.

7. Pipe sleeves must be at least one pipe size larger than the pipe they protect.

8. Toilet compartments must be at least 30 in wide.

9. Relief valves must be installed on water heaters.

10. Toilets are required to have a minimum front clearance of 12 in between the front edge of the toilet and an obstruction.

11. Gang showers are required to be equipped with shower valves that prevent scalding.

12. Cleanout plugs must be made of lead.

13. Plastic closet flanges must have a minimum thickness of ¼ in.

14. The minimum drain size for a shower is 2 in.

15. The minimum size for underground drainage pipe is 2 in.

16. The minimum drain size for a bathtub is 2 in.

17. All horizontal drainage pipe should be supported at 6-ft intervals.

18. A lavatory has a fixture unit rating of 2.

19. A residential toilet has a fixture unit of 4.

20. Hot water is water with a temperature equal to or greater than 100°F.

Test Two

Choose the correct letter for the answer to these questions.

1. Which of the following is a prohibited trap?
   a. P-trap
   b. Deep-seal trap
   c. Crown-vented trap

2. How far can a 2-in trap be from its vent?
   a. 2 ft
   b. 4 ft
   c. 6 ft

3. How many drainage fixture units are kitchen sinks rated for?
   a. 1
   b. 2
   c. 3

4. What is the maximum lead content allowed in solder used with potable water pipe joints?
   a. 0.08 percent
   b. 0.01 percent
   c. 0.02 percent

5. What is the maximum lead content allowed in pipe used to convey potable water?
   a. 8 percent
   b. 1 percent
   c. 2 percent

6. What is the maximum number of sinks or lavatories that may be connected with a continuous waste.
   a. 2
   b. 3
   c. 4

7. What is the minimum size for a floor drain?
   a. 1½ in
   b. 2 in
   c. 3 in

8. What color is the indelible strip on type L copper?
   a. Blue
   b. Red
   c. Green

9. What color is the indelible strip on type M copper?
   a. Blue
   b. Red
   c. Green

10. What color is the indelible stripe on type K copper?
    a. Blue
    b. Red
    c. Green

11. What is the minimum distance allowed horizontally between a vent terminal and a window or ventilating opening in a building?
    a. 5 ft
    b. 10 ft
    c. 15 ft

12. If a vent terminal is too close to a ventilating opening, how far should the vent extend above the opening?
    a. 1 ft
    b. 2 ft
    c. 7 ft

13. What is the minimum clearance that should be in front of a 2-in cleanout?
    a. 6 in
    b. 10 in
    c. 12 in

14. Manholes are not required on which of the following pipe sizes?
    a. 6 in
    b. 10 in
    c. 12 in

15. Which of the following traps is not a trap?
    a. P-trap
    b. Grease trap
    c. Drum trap

Test Three

Answer true or false to each question.

1. An air break can be found in indirect-waste connections.
2. An air gap can be found in indirect-waste connections.
3. The drainage pipe running from outside a building's foundation to the main sewer is called the building drain.
4. Building traps, or house traps, are prohibited.
5. Another name for a clinical sink is bedpan washer.
6. Local vents are used in conjunction with bedpan washers.
7. A hospital must have two water services.
8. Potable water is not safe for drinking, but it is okay for flushing toilets.
9. Vent stack is just another way of saying stack vent.
10. Back-flow preventers are not required on sillcocks.
11. Sewer ejector sumps do not have to be vented.
12. Most buildings containing plumbing must have at least one 3-in main vent.
13. A soil pipe is a pipe that is always buried in the ground.
14. A well is normally considered a private water source.
15. A master's license is the first license most plumbers receive.

Answers to Test One

1. False	11. True
2. False	12. False
3. False	13. True
4. True	14. True
5. False	15. True
6. True	16. False
7. False	17. False
8. True	18. False
9. True	19. True
10. False	20. False

Answers to Test Two

1. c	9. b
2. c	10. c
3. b	11. b
4. c	12. c
5. a	13. c
6. b	14. a
7. b	15. b
8. a	

Answers to Test Three

1. True	9. False
2. True	10. False
3. False	11. False
4. True	12. True
5. True	13. False
6. True	14. True
7. True	15. False
8. False	

Answers to Test One

1. Raise	11. True
2. Raise	12. False
3. Raise	13. True
4. True	14. True
5. False	15. True
6. True	16. False
7. False	17. False
8. True	18. False
9. True	19. True
10. False	20. False

Answers to Test Two

1. d	9. a
2. c	10. c
3. a	11. b
4. c	12. c
5. d	13. a
6. b	14. d
7. b	15. b
8.	

Answers to Test Three

1. True	9. False
2. True	10. False
3. False	11. False
4. True	12. True
5. True	13. False
6. True	14. True
7. True	15. False
8. False	

16

The Conclusion

You have reached the last chapter in this book. By now, you should know more about plumbing than many plumbers do. Since this is the last chapter, I will treat it like the last day of school and keep it short. This chapter rounds out the book and some important facts to remember.

Local Code Variations

I mentioned earlier in the book that local codes vary, but this is an important enough issue to talk about it again. Depending upon where you will be working, the plumbing code may have additional or different requirements from the procedures I have described. For example, I have discussed mobile home parks and recreational parks, but these types of facilities sometimes require special plumbing. To be safe, you must always check and confirm local code requirements before doing any plumbing.

Local codes can change from town to town. I have worked in areas where a 15-min drive would put you in a jurisdiction that uses a different code. With the possibility for so many variables, you must check local requirements in every jurisdiction.

This book has not covered every aspect of all plumbing codes. However, the information here is representative of the most commonly used procedures. Also remember that plumbing codes can be changed, sometimes on short notice. What was acceptable last month may be a code violation this month. The point is, always check for current regulations and requirements before doing any plumbing.

Administration

Chapter 1 dealt with administration. In it, you learned about the paperwork side of the code. As you go through your career, you will not deal with administration matters on a daily basis, but you will find it a factor in your trade.

Regulations, Permits, and Enforcement

Chapter 2 talked about regulations, permits, and enforcement. As a professional plumber, you will be obligated to work with these areas of the codes frequently. I would like to stress a few points on these aspects of the code.

Always apply for a permit when a permit is required. Be sure to remember that a permit is needed for the installation or replacement of water heaters. Never overlook safety requirements. Make it a rule to work with plumbing inspectors, not against them. Do all of your work in compliance with code requirements.

Materials

Chapter 3 was a fairly long chapter on approved materials. In the real world, you will not have a need for all the information in Chap. 3, but for test purposes, you had better know it.

Certainly you will have various materials that you will use on a day-to-day basis. If you favor polyvinyl chloride (PVC) pipe, you are not likely to use acrylonitrile butadien styrene (ABS) pipe, but it does help to know all of the options available to you. The main point relating to materials is to always use a material that is approved for the intended use.

Drainage Systems

Chapter 4 was filled with facts on drainage systems. You may never have a need to install systems for storm drainage. Subsoil drains and roof drains may never become a part of your job description, but drainage is a big part of a plumber's job.

Let's look at some keys to drainage systems. Always use approved materials. Size your pipes to handle the maximum load anticipated. Use approved hangers and support your pipe at frequent intervals. Remember, code requirements are based on minimums; it doesn't hurt to install extra hangers. Never pipe a drainage system to an unapproved location. If you are required to work with chemical wastes or other special wastes, check with local authorities for their current requirements. Install your piping with service in mind. Avoid tight turns and close quarters. Install every job as if it were your own.

Vents

Chapter 5 showed you the importance of vents. Don't take venting lightly. If you learn to use wet-venting techniques, you can save substantial money in labor and material. Just because most of your vents will not carry water doesn't mean they shouldn't be tested for leaks. You have been told about the dangers of sewer gas and depleted trap seals; don't allow faulty pipe joints to create a hazardous circumstance.

Traps and Such

Chapter 6 taught you about traps and such. Always install proper traps on your fixtures. Install an adequate number of cleanouts, and remember to allow enough clearance for the cleanouts to be used. Never cut corners by omitting required interceptors. The issues discussed in Chap. 6 are vital to sanitary plumbing.

Fixtures

Chapter 7 was all about fixtures. Always use approved fixtures. When doing your rough-in, allow for the clearance required around fixtures. If you rough it in wrong, you are looking at the potential for major expenses in the final plumbing. Be careful with fixtures; many of them are fragile and expensive. When bidding jobs, be aware of requirements for handicap fixtures. Omitting required handicap fixtures from your bid will teach you a costly lesson.

Potable Water

Chapter 8 was a very long chapter on potable water. Back-flow prevention is a key issue when installing potable water systems. Proper pipe support and installation is important in water systems. Skimping on supports and air chambers can result in unhappy customers when their pipes start to bang. Pipe sizing is probably the most complicated issue in Chap. 8. Remember, when you are working with potable water, you have the health of the nation in your hands.

Inspections

Chapter 9 told you what you need to know about testing your work and getting it inspected. Don't cheat on testing your work; the results can be disastrous. Be careful with test equipment, especially test balls. Never reach inside a fitting to release the air pressure from a test ball if you are testing with water. The pressure behind the test ball is enough to do very nasty things to your hand and arm if they are caught in the pipe. Always use a long screwdriver or similar tool to release the pressure in test balls used with water tests.

When you are done with your test, remove all your caps and test balls. Many plumbers forget to remove the caps and test balls from vent terminals. If you leave the vent pipes blocked with test equipment, the vents cannot work properly. Remember to protect your pipes with nail plates when necessary. If you don't, leaks resulting from your failure to protect the pipes will be at your expense.

Gas

Chapter 10 revolved around gas work. Your area may not regulate gas work with the plumbing code, but some jurisdictions do. Gas work can be dangerous and potentially lethal. If you are not trained for gas work, don't do it. If you are competent in working with gas, don't make stupid mistakes; there is a lot at stake.

Principles

Chapter 11 preached principles and ethics. The decision for what type of plumber you will be is up to you, but I hope you will be a good one and a respected one.

Graphics

Chapter 12 showed you an array of graphics. You saw tables, charts, and graphs. These graphics represent the type you will find in your code book and on licensing exams.

Licensing

Chapters 13, 14, and 15 strived to help you prepare for licensing requirements. They offered you the benefits of many year's and many people's experience. If you apply these principles, I am sure you will do well on your licensing exam.

The End

Well, we are at the end of the last chapter. This chapter has been a recap, with a few tidbits thrown in for your consideration. I have spent many years in the plumbing trade and many hours writing this book. I hope the combination of my hard-earned experience and my key-stroke-sore fingers will help you along your way to becoming a professional plumber. In return for all I am giving you, I only ask one thing. Please represent the professional plumbing trade proudly. Good luck in all your endeavors.

Glossary

Accessible   Refers, as it relates to plumbing, to a means of access. For example, a tub waste is considered accessible when there is an access panel that can be opened or removed to gain access to the tub waste.

Air break   Refers, in the drainage system, to an indirect-waste procedure. The indirect waste enters a receptor through open air.

Air gap (drainage)   The vertical distance that waste travels through open air, between the waste pipe and the indirect-waste receptor.

Air gap (potable water)   The vertical distance water travels through open air, between the water source and the flood-level rim of its receptor or fixture.

Air gap (the device)   A device used to connect the drainage of a dishwasher with the sanitary drainage system.

Antisiphon   When a device cannot be made to form a siphonic action.

Area drain   A drain used to receive surface water from grounds or parking areas, for example.

Aspirator   A device used to create a vacuum, like in a suction system for medical offices.

Back-flow   The backward flow of water or other liquids in the drainage or water system.

Back-flow preventer   A device used to prevent back-flow.

Back-siphonage   Essentially the same as back-flow; it is the reverse flow of water or liquids in a pipe, caused by siphonic action.

Back-water valve   A device installed on drainage systems to prevent a back-flow from the main sewer into the building where the back-water valve is installed.

Ballcock   An automatic fill device. Ballcocks are most commonly found in toilet tanks. They supply a regulated amount of water, on demand, and then cut off when the water level reaches a desired height.

Branch   A part of the plumbing system that is not a riser, main, or stack.

Branch interval   A means of measurement for vertical waste or soil stacks. A branch interval is equal to each floor level or story in a building, but they are always at least 8 ft in height.

Branch vent Vents that connect individual vents with a vent stack or stack vent. They can be used to connect multiple vents to a vent stack or stack vent.

Building drain The primary drainage pipe inside a building.

Building sewer The pipe extending from the building drain to the main sewer. Building sewers usually begin between 2 and 5 ft outside of a building's foundation.

Building trap A trap installed on the building drain to prevent air from circulating between the building drain and sewer.

Cistern A covered container used to store water, normally nonpotable water.

Code A set of regulations that govern the installation of plumbing.

Code officer An individual responsible for enforcing the code.

Combination waste and vent system A plumbing system in which few vertical vents are used. In these systems, the drainage pipes are often oversized to allow air circulation in the system.

Critical level The point where a vacuum breaker may be submerged, before back-flow can occur.

Cross-connection A connection or situation that may allow the contents of separate pipes to commingle.

Developed length A method of measurement, based on the total liner footage of all pipes and fittings.

Drain A pipe that conveys wastewater or water-borne wastes.

Drainage system All plumbing that carries sewage, rainwater, and other liquid wastes to a disposal site. A drainage system does not include public sewers or sewage treatment and disposal sites.

Existing work Work that was installed prior to the adoption of current code requirements.

Fixture supply The water supply between a fixture and a water-distribution pipe.

Fixture unit A unit of measure assigned to fixtures for both drainage and water, to be used in pipe sizing.

Flood-level rim The point of a fixture where its contents will spill over the rim.

Grade The downward fall of a pipe.

Groundworks Plumbing installed below a finished grade or floor.

Hot water Water with a temperature of 110°F or more.

House trap Same as a building trap.

Interceptor A device used to separate and retain substances not wanted in the sanitary drainage system.

Leader An exterior pipe used to carry storm water from a roof or gutter.

Local vent A vertical vent used with clinical sinks to transport vapors and odors to the outside air.

Main Any primary pipe for water service, distribution, or drainage.

Main sewer The public sewer.

Main vent The primary vent for a plumbing system.

Nonpotable water Water not safe for human consumption.

Offset A change in direction.

Open air The air outside of a building.

Pitch See *Grade*.

Plumbing inspector See *Code officer*.

Potable water Water safe for human consumption in drinking, cooking, and domestic uses.

Private sewage disposal system A sewage disposal system serving a private party, a septic system.

Private water supply A water supply serving a private party, such as a well.

Readily accessible Having direct and immediate access to an object. If an access panel must be removed before the object can be accessed, the object is not readily accessible.

Rim The open edge of a fixture.

Riser A vertical pipe in the water-distribution system that runs vertically for at least one story.

Rough-in The installation of plumbing in areas that will be concealed once the building is completed.

Sewage Liquid waste that contains animal or vegetable matter. The matter may be contained in solution or in suspension and may contain chemicals in solution.

Slope See *Grade*.

Stack A vertical pipe in the drainage system. A stack may be a vent, soil pipe, or waste pipe.

Stack vent A vertical pipe that extends above the highest drainage point to vent the drainage system.

Storm water Rainwater.

Sump vent A vertical vent that rises to vent a sump, such as in a sewer ejector sump.

Tempered water Water tempered to maintain a temperature between 85 and 110°F.

Underground plumbing See *Groundworks*.

Vacuum   A pressure that is less than the pressure produced by the atmosphere.

Vacuum breaker   A device used to prevent a vacuum.

Vent stack   A vertical pipe in the drainage system that acts only as a vent and does not receive the discharge of plumbing fixtures.

Waste   A discharge from fixtures and equipment that does not contain fecal matter.

Water-distribution system   The collection of water pipes within a building that delivers water to fixtures and equipment.

Water-hammer arrestor   A device that defeats water hammer by absorbing pressure surges.

Water main   The public water service.

Water service   The pipe delivering water from a water source to the water-distribution system of a building.

Well   A water source from a hole in the ground.

Wet vent   A pipe that receives the drainage from plumbing and does double duty by venting part of the plumbing system.

Yoke vent   An upward connection from a soil or waste stack to a vent stack. Yoke vents are normally made with wye fittings to prevent pressure changes in the stack.

Index

ABOUT THE AUTHOR

R. Dodge Woodson is a licensed master plumber, plumbing contractor, general contractor, and licensed real estate broker. He is the author of many magazine articles, as well as six books, including *Home Plumbing Illustrated*, and *The Plumbing Apprentice Handbook*, to be published by McGraw-Hill in the Fall of 1993. Mr. Woodson lives in Bowdoinham, Maine. It is to be noted that Woodson is not affiliated with any code agencies.